LIGAND-FIELD PARAMETERS

LIGAND-FIELD PARAMETERS

M. GERLOCH
Assistant Director of Research
University Chemical Laboratory, Cambridge
and Fellow of Trinity Hall

AND

R. C. SLADE
Lecturer in Inorganic Chemistry
Queen Elizabeth College, London

CAMBRIDGE
AT THE UNIVERSITY PRESS
1973

CAMBRIDGE UNIVERSITY PRESS
Cambridge, New York, Melbourne, Madrid, Cape Town, Singapore, São Paulo, Delhi

Cambridge University Press
The Edinburgh Building, Cambridge CB2 8RU, UK

Published in the United States of America by Cambridge University Press, New York

www.cambridge.org
Information on this title: www.cambridge.org/9780521105668

First published 1973
This digitally printed version 2009

A catalogue record for this publication is available from the British Library

Library of Congress Catalogue Card Number: 72–93139

ISBN 978-0-521-20137-7 hardback
ISBN 978-0-521-10566-8 paperback

CONTENTS

throughout the literature. The parameters used vary from coordination number to coordination number, from symmetry to symmetry and from author to author. It is not always clear whether any or all of these parameters are, or can be, related to one another. Nor is it clear how the parameter values deduced from the spectra or magnetism of 'distorted' systems reflect geometrical distortions as opposed to some radial properties which may be related to bonding in some way analogous to the behaviour of $10\,Dq$.

The present book has been written to describe and explore the nature of the purely non-symmetry-determined part of ligand-field theory. Discussion of symmetry properties is only made to introduce and define ligand-field parameters. Accordingly an elementary knowledge of the usual approaches to ligand-field theory is assumed, together with a similar acquaintance with elementary group theory. It is hoped that the subject matter of this book will draw the attention of those already expert in the general area of ligand-field theory. However, the detailed presentation of the material has also been made with final year honours students and young researchers in mind. Some of the ideas described are well-established and some are new. The subject of ligand-field parameters is not closed and to some extent this book presents a progress report which includes commentary of some current areas of disagreement in the literature.

The plan of the book is roughly as follows. The introductory chapter outlines the role of symmetry in ligand-field theory and contrasts it with the function of splitting parameters. Interpretations and predictions of the simplest crystal-field and molecular-orbital approaches to the Spectrochemical Series are reviewed to focus attention on those aspects of ligand-field theory which are not determined by symmetry. Chapter 2 describes the crystal-field formalism, introducing potentials, angular and radial integrals and the multipole expansion. The expansion of the $1/r_{ij}$ operator in terms of spherical harmonics is written in various different ways in order to clarify its use. The significance of $1/r_{ij}$ as a two-electron operator, and hence the fundamental character of the crystal-field model, is elaborated in the third chapter where it is discussed in the more general context of interelectron repulsion parameters. This chapter also presents a simple discussion of interelectron repulsion parameters as determined by group theory and describes some of the approximations involved in their definition. A brief resumé of Trees' correction is included here.

Radial parameters which arise in angularly-distorted systems are described in chapter 4 and the second-order radial parameter Cp is

PREFACE

Most books on ligand-field theory are concerned with the symmetry-determined aspects of the subject. The assignment of d–d spectra and the construction of crystal-field correlation diagrams form the stuff of most conventional texts, whilst the differences between various books in this area usually reflect only different degrees of mathematical sophistication. In the main, the discipline and rigour such books describe refer only to quantities which depend on the angular properties of wavefunctions, radial properties being sequestered into 'proportionality constants'. From an heuristic point of view, such approaches are obviously sensible in that a qualitative understanding of phenomena must always precede a quantitative one. On being confronted with a succession of transition-metal electronic spectra, for example, it is clearly desirable to establish qualitative relationships between them and to assign electronic transitions to experimental bands before commenting upon 'crystal-field strengths'. All this is to say that the usual approach to ligand-field theory is *via* symmetry and group theory. However mathematical, such an approach is essentially qualitative; although a semi-quantitative understanding of spectra and magnetism readily follows.

The principles of crystal-field theory are usually illustrated by reference to systems with cubic symmetry, and Dq as the only scaling parameter: but few transition-metal complexes are exactly octahedral or tetrahedral. In general such distorted molecules display anisotropy in their electronic, or other, properties as evidenced, for example, by magnetic anisotropy and spectral polarization studies of single crystals. Molecules involving coordination numbers other than four or six may be anisotropic without there being distortion from some ideal symmetry: five-coordinate trigonal bipyramidal complexes typify such cases. Less tractable are molecules with rhombic or lower symmetry, though many may be described as axially distorted from an appropriate cubic symmetry precursor. In all cases, departure from cubic symmetry means less information can be had from group theory alone. Crystal-field parameters proliferate in these circumstances, $Dq, Ds, Dt, Cp, D\sigma, D\tau, \rho_2, \rho_4$, being some of the symbols used to label them introduced. Separation of angular distortion from radial parameters is emphasized here and several recent results of magnetic and spectral studies of single crystals are reviewed in this spirit. A similar treatment of D_{4h} symmetry molecules is made in chapter 5 where *Ds* and *Dt* parameters are defined. It is shown how values for these parameters may be recast in terms of *Cp* and *Dq*, thus possibly leading to interesting, and apparently general, trends in the ratio *Cp*/*Dq*. Throughout both chapters, the philosophy of crystal-field parameterization is discussed and the dangers of a too-literal interpretation of the definitions of *Ds, Dt, Cp* and *Dq* are emphasized.

Interpretations of radial parameters begin in chapter 6 which reviews the nature and calculation of 10*Dq*, ranging from the most elementary point-charge model to the latest all-electron, *ab initio*, molecular-orbital calculations. The chapter aims to identify the assumptions and problems of the various methods which have been employed to calculate 10*Dq* and so give an insight into the various factors which really determine this quantity. No attempt is made to provide a basis for actual computation. The difficulties of *ab initio* calculations are such that simpler models must generally be used for understanding splitting parameters for a wide range of compounds and in chapter 7, a fresh appraisal of the unrealistic point-change model is described, largely in a spirit of exploration. Simple trends in the relative behaviour of *Cp* and *Dq* as functions of bond length, effective nuclear charge and ligand charge are deduced which indicate an, albeit temporary, utility for the approach described. Semi-empirical molecular-orbital models, especially the angular overlap method, are alternatives favoured by some authors and these are reviewed in chapter 8. The chapter ends with a comparison of the parameterized point-charge model and the angular overlap method applied to low-symmetry ligand-field parameters.

The Nephelauxetic effect is discussed in chapter 9. The various formalisms used to describe the effect are outlined and the apparently opposing views about evidence for differential orbital expansion are reviewed. The discussion thus centres round parameters conventionally symbolized by $B, C, F_2, F_4, \beta_{33}, \beta_{35}, \beta_{55}$. As in the discussion of the low-symmetry crystal-field parameters, some of the current areas of ignorance and the need for further research are pointed out.

To some extent each chapter may be read independently and in this form may commend itself as a 'teaching review'. All but one chapter end with a listing of particularly relevant and useful texts which are cited by a lower case letter. All other references are cited numerically and listed numerically and alphabetically at the end of the book.

We are most grateful for the time and effort given us by Professors A. D. Buckingham and J. Lewis in numerous discussions and in critically reading much of the manuscript. We should also like to thank Drs D. J. Mackey, E. D. McKenzie, P. N. Quested and W. R. Smail and Professor D. P. Craig for various constructive comments in the earlier stages. Dr Mackey also performed the calculations described in the appendix to chapter 7. Finally we wish to thank Mrs Thora Saunders who greatly simplified the later stages by preparing an accurate typescript so quickly.

July 1972

M. Gerloch
Trinity Hall, Cambridge

R. C. Slade
Queen Elizabeth College
London

1
SYMMETRY VERSUS SPLITTING PARAMETERS

Symmetry plays a central role in the interpretation of transition metal *d–d* spectra. The information contained in these spectra may conveniently be considered under the three headings of splitting patterns, absolute energy differences, and intensities and polarizations. Only absolute energy differences are independent of symmetry. Intensities, though largely determined by non-symmetry factors, do have a qualitative aspect insofar as selection rules, derived from group theory, may be involved. Spectral splitting patterns, especially of highly symmetrical molecules, are very largely symmetry-based features. In a similar way, the orders of magnitude for magnetic moments successfully predicted by the 'spin-only' formula and the principles of crystal-field orbital quenching derive largely from group-theoretical considerations. Even early qualitative estimates of the magnitudes of magnetic anisotropies were made directly from a knowledge of formal ground term degeneracies.

Our understanding of the *magnitudes* of spectral splittings and of the detailed behaviours of magnetic moments and anisotropies, however, rests on theories owing little or nothing directly to symmetry. The size of the spectral splitting factor Dq, for instance, is not determined by group theory. Also, interpretations of the electronic properties of molecules with less than cubic symmetry involve many more such parameters, of which Cp, Dt, Ds, $D\sigma$ and $D\tau$ are perhaps the best known. The nature and use of such quantitative parameters form the subject matter of this book.

The origin and first examples of the use of crystal-field theory concerned magnetic moments and their dependence on the quenching of orbital angular momentum.[e] Thus, while it was realized that lanthanide ions in crystal lattices cannot be completely indifferent to their environment, it was found empirically in the late 1920s that Hund's formula,

$$\mu_{\text{eff}} = g\sqrt{[J(J+1)]}, \tag{1.1}$$

where

$$g = 1 + \frac{J(J+1) - L(L+1) + S(S+1)}{2J(J+1)}, \tag{1.2}$$

[e] References marked with a letter are to be found at the end of the chapter.

a formula derived for free-ions, satisfactorily explained the observed magnetic moments in lanthanide compounds. Crystal-field theory grew out of the discovery that this simplicity did not extend to the magnetic properties of compounds of the main transition block. Van Vleck had derived a complementary formula to (1.1), describing the magnetic moments of free-ions of the first-row transition series where, in first order, spin–orbit coupling effects might reasonably be ignored:

$$\mu_{\text{eff}} = \sqrt{[L(L+1)+4S(S+1)]}. \tag{1.3}$$

Accordingly, the magnetic moments of ions with a Russell–Saunders ^3F ground term, Ni^{2+} for example, should have values of $\sqrt{20} = 4.47$ Bohr magnetons. Experimental values for octahedral compounds of nickel(II), however, are typically *ca.* 3.2 Bohr magnetons. The explanation of these and similar discrepancies came independently in general terms from Stoner and in comprehensive detail from Bethe.[7]

It was recognized that, in compounds, the metal electrons are no longer subject only to the attractive nuclear and the repulsive inter-electron coulombic (and exchange) forces, but also to the influence of neighbouring atoms in the molecule or crystal lattice. Without specifying the nature of this influence, electrostatic or covalent bonding for example, some important conclusions may be made by recourse only to symmetry. Thus we know that eigenstates transforming as F terms in spherical symmetry transform as $A_2 + T_2 + T_1$ terms in cubic symmetry. As discussed below, we also know the relative ordering of the energies of these terms, barring a sign, from symmetry considerations alone. In its most elementary form, crystal-field theory serves to establish this sign by depicting the influencing ligands as negative charges. So it is that crystal-field theory, but mostly group theory, establishes a $^3A_{2g}$ ground term for octahedral nickel(II) compounds, for example. The orbital non-degeneracy of this ground term reduces Van Vleck's formula (1.3) to the well-known 'spin-only' formula,[81]

$$\mu_{\text{eff}} = \sqrt{[4S(S+1)]} = \sqrt{[n(n+2)]}, \tag{1.4}$$

where n is the number of unpaired electrons, and the phrase 'orbital quenching' was coined. We do not make it part of our task to further discuss aspects of crystal-field theory which are the substance of most conventional text-books in the subject: a few are listed at the end of this chapter. While assuming familiarity with these matters, we wish to highlight some important development points in crystal-field theory and to emphasize the differing roles played by group theory on the one hand and 'quantitative' theories on the other.

[7] References marked with a number are to be found on pages 229–32.

The 'spin-only' formula and its embellishments with 'orbital contributions' [which are only a reverse way of describing a situation intermediate between those of (1.3) and (1.4)] go far in explaining room-temperature average magnetic moments. Also based on the symmetry-predicted ground state orbital degeneracies is Van Vleck's early explanation[112] of the sizes of magnetic anisotropies observed in transition-metal complexes. Magnetic anisotropies are concerned with spatial anisotropy in molecules (as opposed to spin anisotropy) and so, in first order at least, with the orbital part of the ground term wavefunctions. A slight departure from cubic symmetry removes the degeneracy of an orbital-triplet or doublet term, different components being associated with different spatial directions in the molecule. In molecules with formal orbital-triplet ground terms, then, unequal thermal population of these components may lead to large magnetic anisotropies. Ions with orbital-singlet ground terms, however, should display no magnetic anisotropy, at least in first order, as distortion has no orbital degeneracy to remove. This purely symmetry-based theory satisfactorily explains, for example the $< 1\,\%$ anisotropy† for high-spin iron(III) compounds with formal ${}^6A_{1g}$ ground terms and the typically 30 % anisotropy for near octahedral ions of cobalt(II) with ${}^4T_{1g}$ ground terms. It is, however, the business of the quantitative side of crystal-field or other theories to explain why these cobalt(II) compounds exhibit 30 % anisotropies rather than, say, 80 %. It is the basis of the quantitative aspects of these theories that we shall be discussing.

We mentioned above how symmetry rules dictate more than just which term may arise for an ion in a molecular or crystal environment. It is instructive to examine this powerful use of symmetry further. The free-ion ground terms in the transition metals are ${}^2D\ {}^3F\ {}^4F\ {}^5D\ {}^6S\ {}^5D\ {}^4F\ {}^3F$ and ${}^2D$ for the d^1 to d^9 configurations, respectively. The group-theoretical transformation rules of lowering the symmetry from spherical to octahedral give

$$D \rightarrow E_g + T_{2g}, \quad F \rightarrow A_{2g} + T_{2g} + T_{1g}, \quad S \rightarrow A_{1g}$$

and are well known. The relative ordering of these terms for the lowest energy Russell–Saunders free-ion terms are shown in figure 1.1. Let us remind ourselves of the basic steps in the argument which allows group theory to derive most of the information in figure 1.1.

In octahedral symmetry a set of five d orbitals splits into a triplet

† Expressed say, as anisotropy of susceptibility versus mean susceptibility, i.e. $\Delta\chi/\bar{\chi}$.

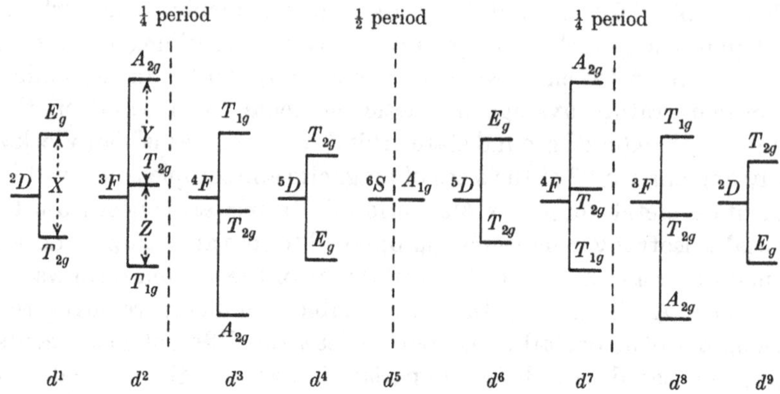

Figure 1.1. Splitting diagrams for d^n configurations in O_h crystal fields, showing inversions in $\frac{1}{2}$ and $\frac{1}{4}$ periods.

and a doublet – the t_{2g} and e_g sets – and these are energetically separated by a quantity we call Δ_{oct} or $10Dq$. In that 'the energy of the t_{2g} set' refers to the energy of a single electron placed in that set, the energy separation of the E_g and T_{2g} *terms* which arise from the d^1 (^2D) configuration is also $10Dq$. The situation for the d^1 configuration is straightforward and well-known. With one exception, all that is involved is group theory. The exception is the *sign* of the orbital or term splittings. In crystal-field terms the metal orbitals are variously repelled by negative ligands and the familiar result in octahedral geometry is that the t_{2g} orbitals lie lower than the e_g orbitals. Insofar as we are concerned with energy *splittings*, a 'baricentre rule' may be invoked such that a d^5 configuration is unshifted energetically on forming a spin-free octahedral complex.[c] Thus, relative to the d^1 configuration, we place the t_{2g} orbitals at $-4Dq$ in energy and the e_g orbitals at $+6Dq$: the same figures pertain for the T_{2g} and E_g terms relative to ^2D.

Conventional crystal-field text-books are extensively concerned with the more complex situation occurring for d configurations involving more than one electron.[b,c] Figure 1.2 shows part of the weak-field/strong-field correlation diagram for octahedral d^2 ions. Only spin-triplet terms are shown. On the left side of figure 1.2 inter-electron repulsion effects split the d^2 configuration into Russell–Saunders terms, Hund's rules leaving ^3F lowest. The nature of these terms, that is their quantum labelling as opposed to their absolute energy separations, is determined entirely by coupling coefficients which in turn derive from the commutation relations between angular

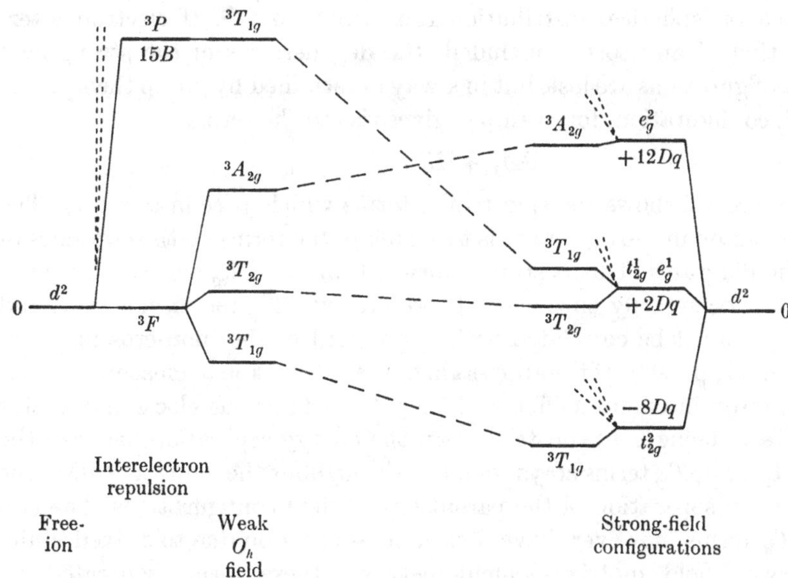

Figure 1.2. Partial correlation diagram for d^2 configuration in O_h crystal field, showing spin-triplet terms only.

momentum operators. Group theory of a slightly different kind tells us that, on lowering the symmetry from free-ion spherical to complex octahedral, the Russell–Saunders terms split into the component terms shown in the figure. At this stage, the ordering and relative energies of the terms arising, say, from ^3F are unknown. As we are considering energy splittings only, the neglect of crystal-field terms possessing full spherical symmetry means that the energy of the $^3T_{1g}(P)$ term should be the same as its parent ^3P term, in first order. Group theory alone has therefore established the qualitative nature of the left-hand side of figure 1.2.

On the right-hand side are represented the strong-field configurations in O_h symmetry, corresponding to placing two electrons in the t_{2g} orbital set, one in t_{2g} and one in e_g, or both in the e_g set. At this strong-field limit, we are supposing that there is no interaction between the pair of metal electrons. Such interaction would involve two-electron operators so that the strong-field representations as one-electron wavefunctions implicitly require the neglect of any electron *interactions*. The ordering of these strong-field configurations derives directly from the ordering of the t_{2g} and e_g orbitals used in the d^1 case: not only the ordering but also the energy values with respect to the free-ion spherical distribution represented by d^2. If electron interaction of any sort is included, the degeneracies of the strong-field configurations are lost, but in a way determined by group theory. The t_{2g}^2 configuration, for example, gives rise to the terms

$$^1A_{1g} + {}^1E_g + {}^1T_{2g} + {}^3T_{1g}.$$

Figure 1.2 shows the spin-triplet terms which arise in this way. The next step in the argument is to correlate the terms on the two sides of the diagrams. This is straightforward for the $^3A_{2g}$ and $^3T_{2g}$ terms as these occur only once each. There are two $^3T_{1g}$ terms, however, and these must be correlated with due regard to the 'non-crossing rule' (ref. 17, p. 200). If the abscissa in figure 1.2 is taken to measure free-ion electron interaction effects, the crystal-field or one-electron operator effects being held constant, then the energy separation between the $^3A_{2g}$ and $^3T_{2g}$ terms stays constant throughout the figure as $10Dq$ – the energy separation of the parent strong-field configurations. The two $^3T_{1g}$ terms, however, have a variable separation due to a fixed, finite crystal-field matrix element between these terms separated by (assumed) variable interelectron repulsions. The calculation of the off-diagonal crystal-field matrix element x may be made without recourse to the specific nature of the crystal-field operator. The process involves the setting up of the $^3T_{1g}$ energy matrix, as follows:

$$\begin{array}{c|cc} & {}^3T_{1g}(F) & {}^3T_{1g}(P) \\ \hline {}^3T_{1g}(F) & -6Dq-E & x \\ {}^3T_{1g}(P) & x & 15B-E \end{array} \tag{1.5}$$

where $15B$ is the $^3F - {}^3P$ energy separation due to interelectron repulsion and E is the energy of either $^3T_{1g}$ term under the combined perturbation of crystal-field and interelectronic repulsions. The diagonal elements are taken with respect to the energy of the ^3F term. In the limit of no interaction between $^3T_{1g}$ terms, the baricentre rule fixes the energy of the $^3T_{1g}(F)$ term as $-6Dq$, relative to ^3F. The interaction between the two $^3T_{1g}$ terms is represented by the off-diagonal element x. This matrix represents the situation anywhere across the abscissa in figure 1.2 as both crystal-field and electron interaction effects are involved. In order to solve this general secular problem, we consider the special case where we know the solutions already; namely, the strong-field limit. Thus, at the right-hand side of the figure, interelectron repulsions are assumed to vanish and hence also B in (1.5). This leads to the quadratic equation:

$$E^2 + 6DqE - x^2 = 0. \tag{1.6}$$

In the strong-field limit, however, we know the roots to be $-8Dq$ and $+2Dq$. Putting E equal to either of these values in (1.6) gives

$$x = \pm 4Dq. \tag{1.7}$$

In summary then, we see how the ordering of the A_{2g}, T_{2g} and T_{1g} terms arising from a free-ion F term and their relative energy separations are determined by group-theoretical considerations. Symmetry even tells us that, relative to X in figure 1.1 being $10Dq$, Y and Z for the d^2 case are $10Dq$ and $8Dq$ respectively. Further, symmetry also relates the splitting patterns for all the d^n configurations, giving rise to the familiar inversions in the half and quarter periods as shown in figure 1.1. These inversion rules may be arrived at *via* the hole formalism. Thus, being concerned here with terms of maximum spin multiplicity, we note that six d electrons will arrange themselves with one electron of one spin per orbital plus a single electron of the other spin. The distribution of this 'extra' electron amongst the five orbitals of 'other spin' will be identical to the distribution of the single electron of the d^1 configuration. The splitting patterns for a d^{5+n} configuration should be identical, therefore, to those for a d^n configuration. In addition, the arrangements of 10-n electrons in the whole d spin-orbital set will be like the arrangements of n electrons but inverted energetically, as arrangements of 10-n electrons are equivalent to arrangements of n holes, i.e. particles of opposite charge. Accordingly, the splitting patterns for d^{10-n} configurations are the inverse of those for d^n. This rule is responsible for the inversion in the half period of figure 1.1 and, in conjunction with the similarity of d^n to d^{5+n}, for the inversions in the quarter period. An alternative way of deriving the d^2-d^3 inversion, for example, would be to set up a correlation diagram like figure 1.2. The left-hand side would be qualitatively identical, in that symmetry alone at first does not appear to determine the relative orderings of A_{2g}, T_{2g} and T_{1g} terms. But, on approaching the problem from the strong-field side, we note that the ordering of strong-field configurations, with the usual single assumption of negatively charged (or dipolar) ligands, is

$$(t_{2g})^3 < (t_{2g})^2(e_g)^1 < (t_{2g})^1(e_g)^2 \text{ etc.}$$

The group-theoretical transformation of these configurations on recognizing electron interactions then determines the ordering of the terms, $^4A_{2g}$ being lowest.

We have reviewed the construction of splitting patterns and correlation diagrams in some detail, partly to provide a basis for later discussions, but mainly to re-emphasize the role of symmetry in these 'crystal-field' arguments. There is one more well-known feature of these diagrams which is neither group-theoretical nor a quantitative feature like Δ_{oct} or $10Dq$. It is the famous $-\frac{4}{9}$ relationship between the theoretical magnitudes of the splitting parameters in octahedral and tetrahedral geometries. At the most elementary level, the inversion implied by the minus sign is conventionally demonstrated by inscribing an octahedron and a tetrahedron in cubes with common axes and examining the relative proximity of d_ϵ (t_2 or t_{2g}) and d_γ(e or e_g) orbital sets to negatively charged ligands. The $\frac{4}{9}$ factor assumes identical natures for the metal and ligands in the two geometries and equal 'bond lengths'. Although the origin of the $-\frac{4}{9}$ factor is not directly group theoretical, it does derive from geometry. It may be seen as due partly to a change in coordination number and partly to the change in size of the cubes in which one inscribes tetrahedra or octahedra with equal bond lengths. As discussed later in this book, the $-\frac{4}{9}$ factor is not a quantitative factor on the same footing as the splitting parameters Dq etc., and as such may be referred to as a geometrical factor.

So we appreciate the strength of simple crystal-field theory. In cubic symmetry, a single parameter is sufficient, in first order at least, to represent the splitting patterns of all d configurations in octahedral, tetrahedral (or other cubic) geometries and to establish orders of magnitude for magnetic properties which are consequential on ground term multiplicities. This power largely derives from group theory. However, when we are interested in absolute energy splittings, in correlations between spectra, magnetism, chemistry and structure, our attention turns to the magnitudes of the splitting factors for which symmetry and group theory have nothing to say.

Perhaps the most celebrated of early discussions on the magnitude of crystal-field splittings was that concerning high- and low-spin iron(III) compounds. Pauling[81] had described his hybridization scheme for octahedral compounds in which six directed covalent bonds could be formed involving overlap of ligand (σ) orbitals with central metal d^2sp^3 hybrids. Six pairs of electrons donated by the ligands fill these directed orbitals. Any further electrons, equal in number to the number of d electrons on the corresponding metal free-ion, had to fit into the remaining, unhybridized d orbitals if possible. So, for iron(III) complexes with the d^5 configuration, six covalent bonds implied five electrons in three d orbitals and hence one unpaired electron. The low magnetic moment of the $Fe(CN)_6^{3-}$ ion was in general agreement with

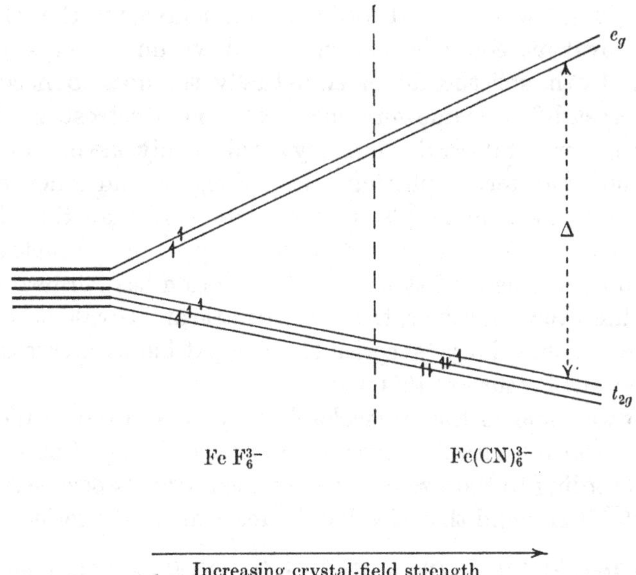

Figure 1.3. Relationship between spin-pairing and Δ in crystal-field model.

this idea. On the other hand, the high moment of FeF_6^{3-} ions, corresponding to five unpaired electrons, implied the non-involvement of the iron 3*d* orbitals in hybridization. Pauling therefore considered the bonding in the complex fluoride to be weaker than in the hexacyanide, more 'ionic' in character and, following the later ideas of Huggins, to involve sp^3d^2 hybrids in which the outer, rather than inner, *d* orbitals participate. In Pauling's view, the change in spin-multiplicity implied a change of bond type.

Van Vleck's[111] approach to the problem was to consider the cyanide ligands as producing a stronger crystal-field than the fluorides, so much so that the ground state involved paired electrons even at the cost of increased interelectron repulsion energy, as in figure 1.3. Pauling[81] found this view unappealing in that fluorine, being the most electronegative element, might be expected to provide a greater electrostatic crystal-field perturbation. However, Van Vleck chose to regard Δ, the splitting factor, as a *parameter* of the system, not necessarily determined solely or largely by purely electrostatic effects. At the same time he showed[111] the relationships between Pauling's hybridization valence-bond method, Bethe's crystal-field model, and Mulliken's molecular-orbital approach,[71] and we shall have more to say about this shortly. The important concept to emerge from the

$FeF_6^{3-}/Fe(CN)_6^{3-}$ work is that all ligand influences, whether electrostatic or covalent, could be parameterized within the crystal-field framework by the splitting factor, Δ. Actually, this does not need to be within a crystal-field framework specifically, i.e. electrostatic, for as discussed earlier, octahedral symmetry requires only one parameter to describe spectral term splittings (neglecting second-order effects involving interelectron repulsion parameters) so that the idea of parameterizing all forms of ligand influence by a single Δ factor is no more than a statement of symmetry. The notion has permeated the chemical literature, however, that electrostatic and covalent bonding effects are somehow both 'surprisingly' compatible with crystal field theory. We shall return to this point also.

And so we come to that remarkable collection of data called the 'Spectrochemical Series'. This is an empirical ordering of metals and ligands according to the size of Δ values their spectra possess.[d] For a given metal it is found that Δ values increase along the series

$$I^- < Br^- < \underline{S}CN^- < Cl^- < F^- < H_2O < \underline{N}CS^- < NH_3 < \text{en} < \text{dipy} < CN^-, \quad (1.8)$$

where en = ethylenediamine and dipy = dipyridyl, and that this series is approximately independent of the central metal ion. A second series, approximately independent of the ligand is:

$$\text{Mn(II)} < \text{Ni(II)} < \text{Co(II)} < \text{V(II)} < \text{Fe(III)} < \text{Cr(III)} < \text{Co(III)} < \text{Ru(III)} < \text{Mo(III)} < \text{Rh(III)} < \text{Pd(IV)} < \text{Ir(III)} < \text{Pt(IV)}. \quad (1.9)$$

The way these two series may be written in terms of only ligands or only metals may be put in the more revealing and remarkable way expressing Δ as a product of a purely ligand function f and a purely metal function g:

$$\Delta \sim f(\text{ligands}).g(\text{metal}). \quad (1.10)$$

Empirical values of f and g, derived from a large number of observed spectra of octahedral compounds are given in table 1.1. There are irregularities, but the values in the table are widely applicable.

One feature of series (1.8) noted earlier, is the way Δ values increase with decreasing size of the donor atom on the ligand:

$$I < Br < Cl < S < F < O < N < C. \quad (1.11)$$

This seems reasonable on the basis of a simple electrostatic origin for the splitting Δ, as smaller ligands imply smaller bond lengths. On a simple electrostatic model, involving no penetration of metal orbitals

TABLE 1.1. *Factorizability of the Spectrochemical Series*

Ligands	f	Metal	$g/10^3\,cm^{-1}$
$6F^-$	0.90	V(II)	12.3
$6H_2O$	1.00	Cr(III)	17.4
6 urea	0.91	Mn(II)	8.0
$6NH_3$	1.25	Mn(IV)	23.0
3 en	1.28	Fe(III)	14.0
$3\ ox^{2-}$	0.98	Co(III)	19.0
$6Cl^-$	0.80	Ni(II)	8.9
$6CN^-$	1.70	Mo(III)	24.0
$6Br^-$	0.76	Rh(III)	27.0
		Re(IV)	35.0
		Ir(III)	32.0
		Pt(IV)	36.0

into the ligand, a charged sphere appears to a point outside to set up an electrostatic potential equal to that if the total charge were at the centre of the sphere. Thus net charge and distance (bond length) are the important factors. What seems less clear from this simple picture is why OH^- and H_2O ligands give very similar crystal-field splittings. Similarly MnF_2 and $Mn(H_2O)_6^{2+}$ have nearly identical spectra. Thus empirically bond lengths appear relevant, as expected from a simple electrostatic model, but net charge does not. These points have been emphasized repeatedly by Jørgensen.[d, 57, 58] He also points out that the increase of Δ values with increasing metal charge, summarized by the empirical inequalities,

$$\Delta \quad \text{for} \quad \mathrm{M(II)} \ll \mathrm{M(III)} < \mathrm{M(IV)}, \tag{1.12}$$

'can hardly be explained by any reasonable electrostatic model'.[d] This statement is based on the idea of increasingly contracted metal orbitals with increased nuclear charge which then interact decreasingly with the negative or dipolar ligands. The further empirical summation of series (1.9), that

$$\Delta \quad \text{for} \quad 3d \ll 4d < 5d \tag{1.13}$$

seems qualitatively compatible with the electrostatic model on the basis of increasing d orbital size, if not of increasing diffuseness.

It is clear, not only from questions stated or implied in earlier paragraphs but also from general chemical knowledge about electrovalent and covalent character, that a purely electrostatic explanation of spectral splitting factors in general and the Spectrochemical Series in particular in unrealistic. The origin of Δ within a *molecular-orbital*

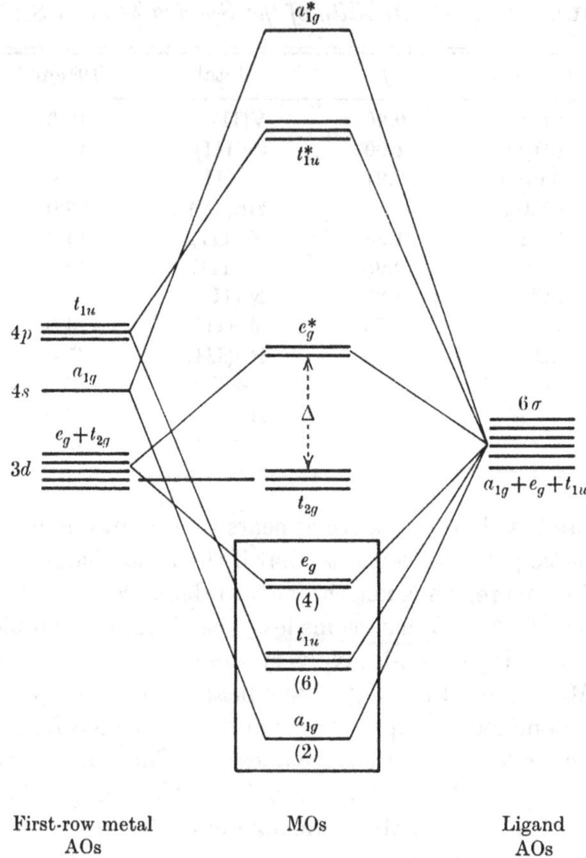

Figure 1.4. Schematic MO diagram for σ-bonding in O_h symmetry.

framework has also been considered for a long time. In very general terms we may look at MO diagrams for octahedral complexes involving first, just σ-bonding and then, with π-bonding also. Schematically the diagram for a σ-bonded octahedral compound is as shown in figure 1.4. Six σ-bonding ligand atomic orbital combinations transform[4] in O_h symmetry as $a_{1g}+t_{1u}+e_g$. These may combine with the metal atomic orbitals transforming as a_{1g}, t_{1u} and e_g for *s*, *p* and *d* orbitals, respectively. No σ-bonding therefore involves the t_{2g} metal orbitals. The diagram, though qualitative, is constructed with the assumption that bonding between ligand and metal *s*, *p* or *d* orbitals are sufficiently similar to leave the e_g^* as the first antibonding orbital above the non-bonding t_{2g} set. Twelve electrons σ-donated by the six ligands fill the lowest three levels (i.e. the six lowest molecular orbitals) whose

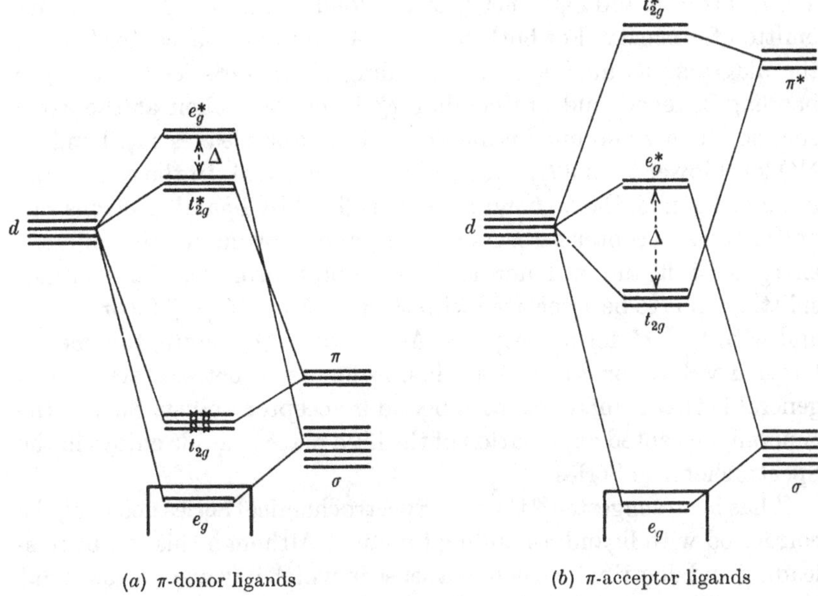

(*a*) π-donor ligands (*b*) π-acceptor ligands

Figure 1.5. Schematic MO diagrams for π-bonding in O_h symmetry.

ordering is therefore unimportant in the present context. These levels are indicated within the box in figure 1.4 and show the analogy with Pauling's valence-bond d^2sp^3 hybrids. The three t_{2g} metal d orbitals are non-bonding both in Pauling's scheme and in the present MO scheme. They house the 'metal' electrons if possible: if not, some electrons go into the antibonding e_g^* level. The non-bonding t_{2g} orbital set with the antibonding e_g^* set above it, thus 'mimic' the t_{2g}–e_g splitting referred to in pure crystal-field theory. The inter-relationship of these three approaches, Pauling's valence-bond method, Bethe's crystal-field approach and Mulliken's molecular-orbital model is clear. In the MO scheme, Δ increases with increased σ-donor 'strength'; i.e. the more the bonding e_g level is depressed in energy, the more the e_g^* antibonding is raised. Δ is associated with the transition $t_{2g} \rightarrow e_g^*$.

It proves convenient to consider the addition of a π-bonding interaction under the separate headings of π-donor and π-acceptor ligands.[f] When discussing π-donor ligands we should consider fairly low-lying filled ligand π-bonding orbitals in contrast to high-lying empty π-antibonding orbitals for π-acceptor ligands. In either case the π-bonding ligand combinations transform as t_{2g} and hence interact with the metal t_{2g} d orbitals. The extreme cases are represented schematically in figure 1.5 in which the low energy σ-bonding a_{1g}

and t_{1u} orbitals and high energy σ-antibonding a_{1g}^* and t_{1u}^* orbitals are omitted for clarity. For both cases of π-donor and π-acceptor ligands the diagrams assume similar σ-bonding characteristics so that the bonding e_g levels and antibonding e_g^* levels are taken at the same energies. The π-bonding interaction in both cases leaves a t_{2g} bonding MO level lower than a t_{2g} antibonding set, as usual. In the case of the π-donor ligands, the t_{2g} bonding level is filled by ligand electrons but not in the π-acceptor case. As a result, and allowing for the different energies of ligand π-donor and π-acceptor orbitals, the 'crystal-splitting' Δ is to be associated with the transition $t_{2g}^* \rightarrow e_g^*$ for π-donors and with $t_{2g} \rightarrow e_g^*$ for π-acceptors. As shown in the figure, this means larger Δ values for π-acceptors than π-donors. A better statement in general is that Δ increases with ligand π-acceptor ability. Such is the generally accepted explanation of the high place cyanide enjoys in the Spectrochemical Series.

It has been suggested[50] that the Spectrochemical Series need only be correlated with ligand π-bonding function. Although this can be misleading and doctrinaire, there is a sense in which it is true. Ideas about the so-called synergic effect of σ-bonding and π-back-bonding are well-known and respected in transition-metal theoretical chemistry (see, for example, ref. *f*). The potential violation of Pauling's electroneutrality principle by increased ligand σ-(or other) donation is offset by the removal of metal negative charge by ligand π-(or other) vacant orbitals. Each mode of bonding facilitates the other to a point of equilibrium. Thus, other things being equal, improved π-acceptor properties in a ligand may be associated with improved σ-donor properties. But as discussed above, increased π-acceptor and σ-donor effects both act together to increase the splitting factor Δ. So, insofar that the effects of ligand π-acceptor and σ-donor properties on Δ are inseparable, one could regard the Spectrochemical Series as governed by one or other alone if one so chose. Such an approach would be at odds with conventional ideas on the variable possibilities of σ-bonding throughout the series of ligands but perhaps not of π-bonding. In any case the separation is indefensible. In tetrahedral symmetry, of course, σ- and π-bonding contributions to Δ are even less separable than in octahedral ones.

There are, then, two main approaches to an understanding of spectral splitting parameters – crystal-field theory with electrostatic potentials, and molecular-orbital theory. It is also well to remember here that there is only one splitting factor in cubic symmetry but many more in molecules of lower symmetry: we are interested in all of them. While we prefer to leave a critical comparison of the two approaches till later, it is convenient here to consider some 'pros and cons'.

Jørgensen has little regard for the utility of crystal-field theory. Considering the *f.g* 'factorizability' of the Spectrochemical Series and other similar matters, he says (ref. *d*, p. 132):

> While all these features justify the greatest optimism regarding the usefulness of ligand-field theory as a *semi-empirical theory*...we must also conclude that the parameter Δ is nearly impossible to predict within a factor of 2 or 5 (as it is impossible to predict most other quantities in theoretical chemistry also) except by the very successful interpolation of empirical series. We may consider Δ as consisting of mainly four contributions:
>
> $$\Delta \sim +\text{electrostatic first-order perturbation} + \sigma(\mathrm{L}\rightarrow\mathrm{M}) - \pi(\mathrm{L}\rightarrow\mathrm{M}) + \pi(\mathrm{L}\leftarrow\mathrm{M}). \quad [1.14]$$

In the preface of his recent book,[57] *Modern Aspects of Ligand-Field Theory*, he is more strident still.

However, crystal-field theory still enjoys much popularity in the introduction of ligand-field theory at elementary levels. This stems partly from the agreement between crystal-field and molecular-orbital theories with regard to the *qualitative* splitting of the *d*-orbital set. It is because much of ligand-field theory is concerned with qualitative questions, such as the role of symmetry, that the apparently simpler electrostatic model has long been retained as a framework within which such questions may be answered. While molecular-orbital theory undoubtedly provides a more satisfying basis by which metal–ligand interactions may be discussed, it does not necessarily follow that molecular-orbital calculations lead to better quantitative predictions. Indeed, the semi-empirical MO methods most favoured by chemists involve gross approximations which lead inevitably to theoretical predictions of questionable significance.

In the quantitative description of the energy levels of metal complexes severe problems exist with both approaches. We can simplify to some extent and say that these problems may be associated with the lack of reality in crystal-field theory and the approximate nature of semi-empirical MO methods. It is pertinent to enquire whether utility is a better judge than reality in deciding between semi-empirical methods.

In the simplest crystal-field model the metal orbitals are supposed to be repelled by the negative part of the ligands which are represented either by point-charges or point-dipoles. There is no penetration of the ligand by the metal orbital and no possibility of electron exchange effects. Objections to the model range from Pauling's comment on $FeF_6^{3-}/Fe(CN)_6^{3-}$ to Jørgensen's remarks on the dependence of Δ on metal oxidation state. The attractions of the model, however, are the simplicity in calculating detailed spectral or magnetic quantities and also the ease in conception. Objections to semi-empirical molecular-orbital models are the increased complexity of calculations and the requirement of such data as valence state ionization potentials or the like, overlap integrals and so on. Further, at the present time, results of calculations on one system (metal, say) do not carry over to others in an obvious way as do those of the crystal-field model. This may change, however. The obvious attraction of MO methods, of course, is their conceptual realism.

The previous paragraph contains many bald statements. We hope to justify our remarks in later chapters. Most work in this area has been done using crystal-field theory and most parameters in magnetochemistry and spectral studies are expressed in these terms. We shall, therefore, study the crystal-field model in detail first and return to molecular-orbital approaches later. At this stage we do not wish to comment further on which approach is 'best'. We are interested in the spectral and magnetic properties of molecules with less than cubic symmetry and hence in the whole array of crystal-field parameters like Cp, Ds and Dt. In discussing these, we develop a 'crystal-field point of view' which seems qualitatively compatible with most experimental observations.

GENERAL REFERENCES

(*a*) L. E. Orgel, ***An Introduction to Transition-Metal Chemistry***, Methuen, London, 1960.

(*b*) B. N. Figgis, ***Introduction to Ligand Fields***, Interscience, New York, 1966.

(*c*) C. J. Ballhausen, ***Introduction to Ligand-Field Theory***, McGraw-Hill, New York, 1962.

(*d*) C. K. Jørgensen, ***Absorption Spectra and Chemical Bonding in Complexes***, Pergamon Press, Oxford, 1962.

(*e*) M. Gerloch and J. Lewis, Chemistry and Paramagnetism, ***Revue de Chimie Minerale***, 1969, **6**, 19.

(*f*) R. S. Nyholm, Structure and Reactivity of Transition Metal Complexes, in ***Proceedings of the Third International Congress on Catalysis***, North Holland, Amsterdam, 1964, p. 25.

2
THE CRYSTAL-FIELD FORMALISM

EXPANSION OF FUNCTIONS

The crystal-field model considers isolated molecules or complex ions in which the central metal electrons are subject to an electric field originating from the surrounding ligands. Earlier papers in the subject used to refer to 'interatomic Stark effects'. The electric field results from a potential V at any point and the effect of V on a metal electron at that point is described quantum mechanically by the energy operator, the Hamiltonian,

$$\mathscr{H}' = -eV. \tag{2.1}$$

Energy calculations of crystal-field splittings require evaluation of matrix elements of the form

$$\int \psi_i^* . eV . \psi_j . d\tau \equiv \langle \psi_i | eV | \psi_j \rangle, \tag{2.2}$$

in which ψs represent metal wavefunctions. In order to proceed we must know how to operate on ψ_j with V. For this we must know the explicit form of the potential operator V. The form of V is established using a most important mathematical theorem called the *expansion theorem*. It is worth while reviewing this theorem, but we do so in an informal way.

Without heed to mathematicians' rigour, the expansion theorem may be stated as follows:

'*Almost any* function of a set of variables may be expressed as a linear combination of a *complete* set of eigenfunctions, of the same variables, of *any* operator.'

We shall comment particularly on the italicized words. Symbolically:

$$f(x_1, x_2, \ldots) = a_1 \phi_1(x_1, x_2, \ldots) + a_2 \phi_2(x_1, x_2, \ldots) + \ldots + a_n \phi_n(x_1, x_2, \ldots) + \ldots, \tag{2.3}$$

where f is a function of the variables $(x_1, x_2, \ldots)$, expanded in terms of the ϕs, with expansion coefficients a_i, where the ϕs form a complete set of eigenfunctions of some operator Ω:

$$\Omega \phi_i = \lambda_i \phi_i. \tag{2.4}$$

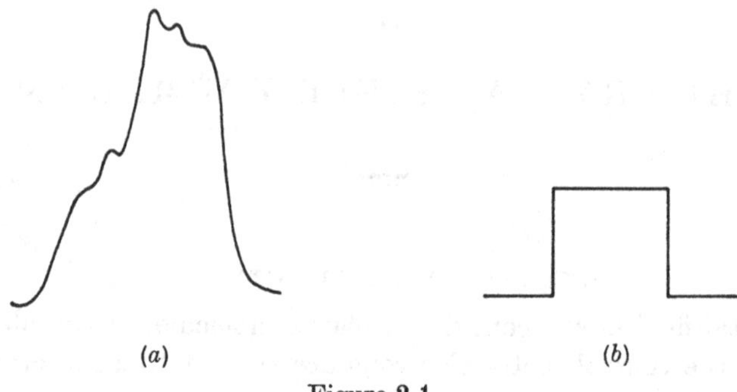

Figure 2.1.

Consider the function of a single variable x. This may look as unappealing as those in figure 2.1. There are limitations on the sorts of functions which may be expanded as in (2.3) but we should be all right if we avoid those which are not single-valued, have infinite discontinuities or, over the variable range to be expanded, have an infinite number of finite discontinuities.

We may choose *any* operator according to the theorem, let us choose $\partial^2/\partial x^2$. Eigenfunctions of an operator are those which satisfy (2.4) in that they are unchanged by the operator except for a numerical multiplicative factor. For $\partial^2/\partial x^2$ we could choose the set of functions $\sin nx$:

$$\frac{\partial^2}{\partial x^2}(\sin nx) = -n^2(\sin nx). \tag{2.5}$$

The proper definition of 'completeness' is a complicated matter but for our purposes it means all possible variations of the type of function (here all n in $\sin nx$) which form an orthogonal set, i.e. are independent of one another. In the present example, we take $\sin nx$ for all positive integer n. According to the expansion theorem, therefore, we may write a general function of x as:

$$\begin{aligned} f(x) &= a_1 \sin x + a_2 \sin 2x + \ldots + a_n \sin nx + \ldots \\ &= \sum_k^\infty a_k \sin kx. \end{aligned} \tag{2.6}$$

Notice that the complete set is infinitely large. We could equally well have taken the eigenfunctions of $\partial^2/\partial x^2$ as $\cos nx$ or indeed the combination $(\cos nx + i \sin nx) \equiv e^{inx}$. These are all examples of Fourier series which are thus special cases of the expansion theorem. In practice it is helpful to choose a basis set, i.e. a set of eigenfunctions, such that the expansion coefficients a_k tend to vanish as n increases giving a (rapidly) convergent series. Such is the hope, for example, in perturbation theory. It is worth remembering, however, that this is a matter of convenience rather than of necessity.

Another way of looking at the expansion theorem is by noting the close mathematical analogy between quantum-mechanical functions and vectors.[27] We may write a vector A as a linear combination of the base vectors x_1, x_2, x_3:

$$A = a_1 x_1 + a_2 x_2 + a_3 x_3 = \sum_k^3 a_k x_k. \qquad (2.7)$$

Note the independence, or orthogonality, of the base vectors x_k. The coefficients a_k tell us how much of each base vector is in A. They are the projections of A on the base vectors. The dot products

$$a_k = x_k . A, \qquad (2.8)$$

and so (2.7) may be written:

$$A = \sum_k^3 x_k(x_k . A). \qquad (2.9)$$

We could equally well express A in terms of the base vectors y_j: these will be related to the base vectors x_k by some transformation – a rotation perhaps:

$$A = \sum_j^3 y_j(y_j . A). \qquad (2.10)$$

The expansion coefficients $\quad b_j = (y_j . A) \qquad (2.11)$

are now different as we have expanded A in terms of a different set of basis vectors. All this has a formal analogy in the expansion of functions. Writing

$$f(x) = a_1\phi_1(x) + a_2\phi_2(x) + \ldots + a_n\phi_n(x) + \ldots \qquad (2.12)$$

we have expressed the function $f(x)$ in terms of the basis functions $\phi(x)$. We could equally well expand them in terms of another basis, e.g.

$$f(x) = b_1\psi_1(x) + b_2\psi_2(x) + \ldots + b_n\psi_n(x) + \ldots. \qquad (2.13)$$

The bases must be complete. In the case of vectors in three-dimensional space we require three base vectors. There may be many more, even an infinite number, in the many-dimensional Hilbert space, as it is called, in which we choose to expand $f(x)$. The analogy between vectors and functions extends also to the coefficients a_k or b_j. Thus if we premultiply (2.12) by $\phi_n^*(x)$ and integrate, we get

$$\int \phi_n^*(x) . f(x) . dx = a_1 \int \phi_n^*(x) . \phi_1(x) . dx + \ldots + a_n \int \phi_n^*(x) . \phi_n(x) . dx + \ldots \qquad (2.14)$$

so that, by orthogonality of the basis functions ϕ,

$$a_n = \int \phi_n^*(x) . f(x) . dx \equiv \langle \phi_n | f \rangle. \tag{2.15}$$

The integral $\langle \phi_n | f \rangle$ corresponds to the dot product of the vectors x_k and A in (2.8): $\langle \phi_n | f \rangle$ tells us how much of $\phi_n(x)$ is in $f(x)$. From (2.12) and (2.15) we have:

$$f(x) = \sum_k^n |\phi_k\rangle \langle \phi_k | f \rangle. \tag{2.16}$$

There are, of course, many more facets of the vector-function analogy. While they are fascinating, we do not require them for the present task. We have presented this discussion in order to give two ways of looking at the expansion theorem. It is a discussion which might well precede an exposition of perturbation theory or of the variational principle. The expansion theorem is used throughout quantum mechanics. We wish to use it now to determine the crystal-field potential eV.

THE CRYSTAL-FIELD POTENTIAL AS AN EXPANSION

The potential eV set up by an arrangement of ligands around a central metal immediately presents us with a problem. We wish to operate with eV, which is produced by ligands at various origins in space, on metal electronic wavefunctions which are referred to a single origin – the central metal ion. It is therefore convenient (though not necessary) to express eV in a form referred to the same single origin. The usual way of doing this is to expand the potential eV as a series of spherical harmonics centred on the metal ion. Thus we might normally regard a ligand as a point-charge or collection of point charges and so express the electrostatic potential it sets up as inversely proportional to the distance of the charges. Instead we express the potential, or rather the sum of potentials from all the ligands, in terms of a coordinate frame whose origin lies elsewhere – on the metal ion. That we can do this follows from the expansion theorem. The potential occurs in the three spatial dimensions and so we require three coordinates for the basis functions in our expansion. It is particularly convenient, though again not essential, to deal with polar coordinates and express V as:

$$V = \sum_k \sum_q c_k^q . R_k(r) . Y_k^q(\theta, \phi). \tag{2.17}$$

By analogy with (2.12), c_k^qs are expansion coefficients of V with respect to the basis functions $R_k(r) Y_k^q(\theta, \phi)$. Note that the basis functions involve all three coordinates r, θ, ϕ even though we have separated them as spherical harmonics $Y_k^q(\theta, \phi)$ and radial functions $R_k(r)$. Note also that at this point we need specify nothing about the radial functions $R_k(r)$. To be quite clear, we write the early part of (2.17) *in extenso*:

$$V = c_0^0 R_0 Y_0^0 + c_1^1 R_1 Y_1^1 + c_1^0 R_1 Y_1^0 + c_1^{-1} R_1 Y_1^{-1} + c_2^2 R_2 Y_2^2 + \dots \quad (2.18)$$

Clearly (2.17) includes harmonics from Y_0^0 to Y_∞^∞ and as such would be comparatively useless: but symmetry and group theory are able to impose an early series termination.

The power of group theory is only realized in the present problem when the operator is viewed in the context in which it is to be used. Matrix elements are the observable quantities, not operators. The matrix elements we must evaluate, in (2.2), may be factorized into radial and angular parts. Let $R(\psi)$ and $A(\psi)$ stand for the radial and angular parts of ψ: similarly $R(V)$ and $A(V)$ for the radial and angular parts of V. Then a typical matrix element in the series is

$$\begin{aligned} M &= \langle \psi_i | V | \psi_j \rangle \\ &= \langle R(\psi_i) | R(V) | R(\psi_j) \rangle . \langle A(\psi_i) | A(V) | A(\psi_j) \rangle . \end{aligned} \quad (2.19)$$

Substituting (2.18) we have:

$$\begin{aligned} M = {} & c_0^0 \langle R(\psi_i) | R_0 | R(\psi_j) \rangle \langle A(\psi_i) | Y_0^0 | A(\psi_j) \rangle \\ & + c_1^1 \langle R(\psi_i) | R_1 | R(\psi_j) \rangle \langle A(\psi_i) | Y_1^1 | A(\psi_j) \rangle + \dots \\ & + c_k^q \langle R(\psi_i) | R_k | R(\psi_j) \rangle \langle A(\psi_i) | Y_k^q | A(\psi_j) \rangle + \dots . \end{aligned} \quad (2.20)$$

Now we have not specified the radial parts of either the potential or the metal wavefunctions so we cannot expect to say anything about the radial integrals

$$\langle R(\psi_i) | R_k | R(\psi_j) \rangle .$$

But the angular integrals

$$\langle A(\psi_i) | Y_k^q | A(\psi_j) \rangle$$

are specified. If in our particular problem any of these angular integrals vanish identically, i.e. by virtue of symmetry, then there is no point in knowing the associated radial integral and more important, no point in knowing the associated expansion coefficient c_k^q. In other words, by looking at the group-theoretical behaviour of the angular parts of the matrix elements of V we can hope to identify those terms in the expansion of V which are relevant.

SERIES TERMINATORS

A most useful series terminator is derived from the specification of the wavefunctions ψ. If ψ_i, ψ_j are metal d orbitals, the angular matrix elements of (2.20) vanish for Y_k^q with $k > 4$; if metal f orbitals, they vanish for $k > 6$. Let us see why. One of the advantages of expanding V in terms of spherical harmonics is that the metal wavefunctions on which they operate are conventionally expressed in the same way. Without being too specific for the moment about the detailed radial parts of these wavefunctions, we can write

$$\psi \sim R(\psi).Y_l^m = R(\psi).\Theta_l^m.\Phi_m. \tag{2.21}$$

For d orbitals, for example, the angular parts of the wavefunction involve the spherical harmonics Y_2^m. The matrix elements in (2.20) therefore involve angular integrals of the form:

$$\langle Y_{l'}^{m'}|Y_k^q|Y_l^m\rangle. \tag{2.22}$$

We use the letters k and q to label l and m quantum numbers of the harmonics appearing in the potential. We should note a further detail here. The Y_l^ms in (2.22) refer to *orbitals*, not terms. This is because the crystal-field operator V is a one-electron operator and so for operation on a many-electron wavefunction, that is a term, it must be expressed as a sum of one-electron operators, e.g. for d^2

$$V = V(1)+V(2), \tag{2.23}$$

in which $V(1)$ operates only on electron (1), $V(2)$ only on electron (2). We shall discuss this matter later, but for the moment we note the first step† in evaluating matrix elements of the form

$$\langle L, M_L|V|L', M'_L\rangle$$

is to expand them as $\langle l, m|V|l', m'\rangle$,

as in (2.22).

The matrix elements (2.22) vanish unless a vector triangle can be constructed from the three suffices l', k and l: if $l = l' = 2$ for d orbitals, then $k \leqslant 4$, for example. The proof of this is simple and educational. Consider the second half of a matrix element like (2.22),

$$Y_k^q|Y_l^m\rangle, \tag{2.24}$$

Y_k^q operates on $|Y_l^m\rangle$. This may be considered as the coupling together of Y_k with Y_l and so, according to the usual vector coupling rules, vectors of lengths $k+l$, $k+l-1$, ..., $k-l$ are obtained: remember k and l are the vector moduli and q and m their orientations, i.e. their z-projections.

† This expansion need not involve an *explicit* breakdown of $|L, M_L\rangle$ terms into $|l, m_l\rangle$ orbital determinants, as the process is determined by the commutation properties of the angular momentum operators. The use of Wigner coefficients or $3j$ symbols and the general machinery of tensor operators deals with such processes *implicitly*. We do not discuss these powerful *techniques* in a book devoted to *principles* (see, for example, ref. 8).

THE VECTOR TRIANGLE RULE[56]

In parenthesis let us consider why we should regard (2.24) as a coupling of angular momenta. In spherical symmetry we have representations of order 1, 3, 5, ..., $(2l+1)$ corresponding, for example, to the s, p, d... orbital sets as bases. Let us write such representations as $\mathscr{D}_0, \mathscr{D}_1, \mathscr{D}_2, \ldots, \mathscr{D}_n$. Suppose a spherical atom or ion has two electrons which respectively belong to the representations $\mathscr{D}_k$ and $\mathscr{D}_l$ (say a p electron and a d electron). What representations of the spherical group do these electrons *as a pair* span? That is, we are considering the product wavefunctions of the type, $\psi_i(1).\psi_j(2)$ etc., or $Y_l^m(1).Y_{l'}^{m'}(2)$. To answer the question we take the direct product of the individual electron representations and reduce it. Thus we want to know the coefficients a_i in

$$\mathscr{D}_k \wedge \mathscr{D}_l = \sum_i a_i \mathscr{D}_i. \tag{2.25}$$

We use the well-known result[d] that the character of the direct product representation equals the product of the individual representation characters,

$$\chi(\mathscr{D}_l \wedge \mathscr{D}_k) = \chi(\mathscr{D}_l).\chi(\mathscr{D}_k). \tag{2.26}$$

Now consider electronic wavefunctions transforming in the spherical group in the representation $\mathscr{D}_l$. These have the form $\Theta_l^m.e^{im\phi}$. The transformation matrix for these functions under a rotation α will be the representation $\mathscr{D}_l$. Thus

$$\mathscr{D}_l = \begin{pmatrix} e^{il\alpha} & & 0 \\ & e^{i(l-1)\alpha} & \\ & \ddots & \\ 0 & & e^{-il\alpha} \end{pmatrix}, \tag{2.27}$$

and the character of $\mathscr{D}_l$ is the sum

$$\chi(\mathscr{D}_l) = (e^{il\alpha} + e^{i(l-1)\alpha} + \ldots + e^{-il\alpha}). \tag{2.28}$$

On substitution in (2.26) we find:

$$\begin{aligned} \chi(\mathscr{D}_k \wedge \mathscr{D}_l) &= (e^{ik\alpha} + e^{i(k-1)\alpha} + \ldots + e^{-ik\alpha})(e^{il\alpha} + e^{i(l-1)\alpha} + \ldots + e^{-il\alpha}) \\ &= (e^{i(k+l)\alpha} + e^{i(k+l-1)\alpha} + \ldots + e^{-i(k+l)\alpha}) \\ &\quad + (e^{i(k+l-1)\alpha} + e^{i(k+l-2)\alpha} + \ldots + e^{-i(k+l-1)\alpha}) \\ &\quad + \ldots + (e^{i(|k-l|)\alpha} + e^{i(|k-l|-1)\alpha} + \ldots + e^{-i(|k-l|)\alpha}) \\ &= \chi(\mathscr{D}_{k+l}) + \chi(\mathscr{D}_{k+l-1}) + \ldots + \chi(\mathscr{D}_{|k-l|}). \end{aligned} \tag{2.29}$$

That is:

$$\mathscr{D}_k \wedge \mathscr{D}_l = \mathscr{D}_{k+l} + \mathscr{D}_{k+l-1} + \ldots + \mathscr{D}_{|k-l|}. \tag{2.30}$$

Equation (2.30) shows that the product wavefunctions span representations with suffices ranging from the sum to the difference of the suffices of the original representations. If we had been considering two d wavefunction sets, then (2.30) tells us that the product wavefunctions in the $(nd)^1.(md)^1$ configuration span representations $\mathscr{D}_4$, $\mathscr{D}_3$, $\mathscr{D}_2$, $\mathscr{D}_1$ and $\mathscr{D}_0$, i.e. 9, 7, 5, 3, 1-fold representations, bases for which we call terms G, F, D, P and S. The proof is general, however, for the 'coupling' of any bases represented with respect to the angular momentum operators.

So returning to matrix elements of the type (2.22)

$$M = \langle Y_{l'}^{m'} | Y_k^q | Y_l^m \rangle, \tag{2.22}$$

we couple any two together (*any* two because the spherical harmonics commute) and deduce the span of the product. $Y_k^q | Y_l^m \rangle$ gives rise to harmonics Y_L^M in which L takes all values from $k+l$ to $|k-l|$. Now in order for (2.22) to be non-zero, l' must lie in the range $k+l$ to $|k-l|$ because of the orthogonality of spherical harmonics, *viz.*

$$M = c_1 \langle Y_{l'}^{m'} | Y_{k+l} \rangle + c_2 \langle Y_{l'}^{m'} | Y_{k+l-1} \rangle + \ldots + c_n \langle Y_{l'}^{m'} | Y_{|k-l|} \rangle. \tag{2.31}$$

The individual matrix elements in (2.31) may vanish because of orthogonality in the m quantum numbers, but we can at least demand that l', k and l in (2.22) form a 'vector triangle'. In this way we know that, for crystal-field matrix elements between d orbitals, terms in the potential V of order (k) greater than 4 are inoperative.

The second generally useful series terminator for (2.20) derives from the parity of the three functions in $\langle \psi_i | Y_k^q | \psi_j \rangle$. The general rule is that the triple product of the functions under the integral signs in these matrix elements must be of even parity. We are discussing here the inversion operator which inverts only the electrons so that the above parity rule remains pertinent even for molecules which lack a centre of symmetry. For example, if ψ_i and ψ_j are both d orbitals, Y_k^q must have k even for the matrix element not to vanish and this is true regardless of the symmetry of the molecule e.g. octahedral with a centre of inversion or tetrahedral, without one.

Thus far, by noting the spherical basis of the functions we use and their parity, the series in (2.20) is greatly shortened. For matrix elements of d electrons, the effective potential is reduced to terms involving Y_0^0, Y_2^q and Y_4^q. Further limitations are imposed by the requirements of the ligand arrangement round the metal ion, that is by the site symmetry. We shall consider the case for tetrahedral symmetry, T_d. We required eV to transform as the totally symmetric representation Γ_1 as it is part of the total Hamiltonian operator (and the free-ion Hamiltonian itself transforms as Γ_1) and we know that the energy of a molecule cannot depend on how we look at it. Y_0^0 obviously occurs in potentials for intra-subshell matrix elements (i.e. excluding $\langle p|V|d\rangle$ etc.) in any group as Y_0^0 is spherically symmetric and transforms as Γ_1. But, because it is spherically symmetric, the potential it represents cannot give rise to orbital *splittings* and is only concerned with overall energy shifts. As we generally take our basis sets from the same subshell, e.g. $3d$ functions for the first-row transition block, uniform radial parameters should prevail and all terms arising from that subshell should suffer energy shifts of identical magnitudes. Spectra (and magnetism to some extent) are concerned with energy *differences* and so we do not consider the spherically symmetric term Y_0^0 further. When we refer to the tetrahedral potential, or whatever, later, we mean that part responsible for orbital splitting.

THE TETRAHEDRAL POTENTIAL FROM SYMMETRY[b]

Returning to the crystal-field potential in T_d symmetry, we consider next the transformation properties of the harmonic set Y_2^m. Using the well-known formula for the character of a spherical representation under a rotation α (ref. *d*, p. 192)

$$\chi(\alpha) = \frac{\sin(l+\frac{1}{2})\alpha}{\sin(\frac{1}{2}\alpha)}, \tag{2.32}$$

we find (recalling that $\hat{\sigma} \equiv \hat{C}_2 . \hat{\imath}$ and $\hat{S}_4 \equiv \hat{C}_4 . \hat{\imath}$ in this context and that the even-order harmonics are even under the inversion operation $\hat{\imath}$) the following characters under the operations of T_d:

	E	$8C_3$	$3C_2$	$6S_4$	$6\sigma_d$	
$\chi(l=2)$	5	-1	1	-1	1	(2.33)

and the irreducible components of Y_2^m in T_d are $E+T_2$. The totally symmetric representation is not involved and so Y_2^ms cannot appear in the potential eV.

For the set Y_4^m we have the table:

	E	$8C_3$	$3C_2$	$6S_4$	$6\sigma_d$	
$\chi(l=4)$	9	0	1	1	1	(2.34)

giving the irreducible components of Y_4^m in $V(T_d)$ as

$$A_1 + E + T_1 + T_2.$$

Thus one linear combination of the Y_4^ms, transforming as A_1 occurs in the potential. We may determine which, by specific investigation of the transformation properties of the individual spherical harmonics in T_d symmetry. We consider here the case in which the system is quantized along the four-fold S_4 axis; that is, S_4 corresponds to the z-axis to which the spherical harmonics are referred. In this case, rotations about axes parallel to z only affect the Φ function, transforming $e^{im\phi}$ into $e^{im(\phi+\alpha)}$ for rotations of α. Hence:

$$\hat{C}_4 . e^{4i\phi} \to e^{4i(\phi+\frac{1}{2}\pi)} \to e^{4i\phi} . 1$$

$$\hat{C}_4 . e^{3i\phi} \to e^{3i(\phi+\frac{1}{2}\pi)} \to e^{3i\phi} . i \quad \text{etc.}$$

so that†

$$\hat{C}_4 . \begin{bmatrix} Y_4^4 \\ Y_4^3 \\ Y_4^2 \\ Y_4^1 \\ Y_4^0 \\ Y_4^{-1} \\ Y_4^{-2} \\ Y_4^{-3} \\ Y_4^{-4} \end{bmatrix} \to \begin{bmatrix} Y_4^4 \\ iY_4^3 \\ -Y_4^2 \\ iY_4^1 \\ Y_4^0 \\ -iY_4^{-1} \\ -Y_4^{-2} \\ -iY_4^{-3} \\ Y_4^{-4} \end{bmatrix} \tag{2.35}$$

and the same is true for $\hat{S}_4$ because l is even. Now we require V to transform as A_1 and so $\hat{S}_4 . V = V$ which can only be true if

$$V_{T_d} = aY_4^4 + bY_4^0 + cY_4^{-4}. \tag{2.36}$$

The next step is easier if we write the harmonics in (2.36) in their cartesian form, *viz.*

$$\left.\begin{aligned} Y_4^4 &= \left(\sqrt{\frac{9}{4\pi}}\right)\left(\sqrt{\frac{35}{128}}\right)\left(\frac{x+iy}{r}\right)^4, \\ Y_4^0 &= \left(\sqrt{\frac{9}{4\pi}}\right)\left(\sqrt{\frac{1}{64}}\right)\left(\frac{35z^4-30z^2r^2+3r^4}{r^4}\right), \\ Y_4^{-4} &= \left(\sqrt{\frac{9}{4\pi}}\right)\left(\sqrt{\frac{35}{128}}\right)\left(\frac{x-iy}{r}\right)^4. \end{aligned}\right\} \tag{2.37}$$

We examine the transformation properties of these three harmonics under $\hat{\sigma}_d$. As the even-order harmonics are unchanged by the inversion operation we need only consider the operation $\hat{C}_2$: this is about a two-fold axis perpendicular to z. Hence if C_2 is taken parallel to x:

$$\hat{C}_2 . \begin{pmatrix} x \\ y \\ z \end{pmatrix} \to \begin{pmatrix} x \\ -y \\ -z \end{pmatrix}, \tag{2.38}$$

† Condon–Shortley phase ignored here, cf. ref. *b*.

and so, from (2.37), we have:

$$\hat{C}_2\begin{pmatrix} Y_4^4 \\ Y_4^0 \\ Y_4^{-4} \end{pmatrix} \rightarrow \begin{pmatrix} Y_4^{-4} \\ Y_4^0 \\ Y_4^4 \end{pmatrix}, \tag{2.39}$$

from which $a = c$ in (2.36) as the trace under C_2 must be unity. Ignoring normalization we then have

$$V_{T_d} = Y_4^0 + d(Y_4^4 + Y_4^{-4}). \tag{2.40}$$

Under $\hat{C}_3$, the cartesian coordinates interchange cyclically in the present quantization scheme:

$$\hat{C}_3\begin{pmatrix} x \\ y \\ z \end{pmatrix} \rightarrow \begin{pmatrix} y \\ z \\ x \end{pmatrix}. \tag{2.41}$$

We require $\hat{C}_3 V = V$ for invariance of the Hamiltonian, so that

$$Y_4^0 + d(Y_4^4 + Y_4^{-4}) = \hat{C}_3[Y_4^0 + d(Y_4^4 + Y_4^{-4})],$$

i.e.

$$\tfrac{1}{8}(35z^4 - 30z^2r^2 + 3r^4) + d\left(\sqrt{\frac{35}{128}}\right)[(x+iy)^4 + (x-iy)^4]$$

$$= \tfrac{1}{8}(35x^4 - 30x^2r^2 + 3r^4) + d\left(\sqrt{\frac{35}{128}}\right)[(y+iz)^4 + (y-iz)^4]. \tag{2.42}$$

Collecting terms in z^4:

$$\tfrac{3}{8} + 2d\left(\sqrt{\frac{35}{128}}\right) = 1, \quad \text{therefore} \quad d = \sqrt{\frac{5}{14}}. \tag{2.43}$$

Hence the potential V_{T_d} transforms as:

$$V_{T_d} = Y_4^0 + \left(\sqrt{\tfrac{5}{14}}\right)(Y_4^4 + Y_4^{-4}). \tag{2.44}$$

Notice that according to (2.17) each harmonic should be multiplied by a function of the coordinate r. As each term in (2.44) is multiplied by the same $R_4(r)$ and as there are no symmetry operations in this (or any other) point group which change scale, i.e. symmetry operations leave $R(r)$ invariant, the $R_4(r)$ is taken as a common factor outside V_{T_d}. In other words, our definition of V has changed, in that we need specify only the non-spherical part, that is the angular properties of V. Of course the size of matrix elements $\langle\psi_i|eV|\psi_j\rangle$ will depend on the function $R(r)$ but we can proceed quite far without knowing this function.

Let us be clear at this point. We have assumed no more than that the ligand influence is electrostatic in nature which means the metal electrons are subject to a *potential*: in consequence we are using a one-electron operator. This important assumption of crystal-field theory is the only one so far used. Group theory and the restriction to a consideration of d orbital splittings then dictated the final form of V as in (2.44). Let us now see what else we can deduce from this single assumption that we are dealing with a potential in tetrahedral symmetry.

MATRIX ELEMENTS[c]

We shall calculate the relationships between the matrix elements of a d orbital set under V_{T_d}. We already know the forms of the angular parts of the d wavefunctions, for these are determined solely by group-theoretical requirements of spherical symmetry: they are the spherical harmonics. We do not know the form of the radial parts of the metal orbitals and at this stage we shall assume nothing about them, simply writing R_{nl}. So for the d orbitals with m_l quantum numbers $2 \rightarrow -2$, we have:

$$\left.\begin{aligned} d_0 &= R_{nl}\left(\sqrt{\frac{1}{2\pi}}\right)\left(\sqrt{\frac{5}{8}}\right)(3\cos^2\theta - 1), \\ d_{\pm 1} &= \mp R_{nl}\left(\sqrt{\frac{1}{2\pi}}\right)\left(\sqrt{\frac{15}{4}}\right)\cos\theta\sin\theta . e^{\pm i\phi}, \\ d_{\pm 2} &= R_{nl}\left(\sqrt{\frac{1}{2\pi}}\right)\left(\sqrt{\frac{15}{16}}\right)\sin^2\theta . e^{\pm 2i\phi}. \end{aligned}\right\} \tag{2.45}$$

The coefficients in (2.45) ensure normalization; normalization to unity in this case, although that is not essential. We wish to evaluate integrals of the form

$$\langle\psi_i|V|\psi_j\rangle = \langle Y_l^{m'}|Y_k^q|Y_l^m\rangle\langle R_{nl}|R(V)|R_{nl}\rangle. \tag{2.46}$$

The radial integral is the same for all matrix elements, that is for all m, m' and q, for constant k and l. Therefore we look at the relationships† between the integrals $\langle Y_l^{m'}|Y_k^q|Y_l^m\rangle$.

These matrix elements may be factorized into two separate integrals involving the coordinates θ and ϕ, *viz.*

$$\langle Y_l^{m'}|Y_k^q|Y_l^m\rangle = \langle\Theta_l^{m'}|\Theta_k^q|\Theta_l^m\rangle\langle\Phi_{m'}|\Phi_q|\Phi_m\rangle. \tag{2.47}$$

Let us consider first the Φ integrals. These can take the more explicit form

$$\int_0^{2\pi}(e^{im'\phi})^* . e^{iq\phi} . e^{im\phi} . d\phi, \tag{2.48}$$

† The Wigner–Eckart theorem tells us that the integrals $\langle Y_l^{m'}|Y_k^q|Y_l^m\rangle$ can be expressed as the product of a part dependent on m etc. and a part independent of m. The independent part, involves a reduced matrix element $\langle Y_l\|Y_k\|Y_l\rangle$. The m-dependent parts are related to one another *via* the Wigner coupling coefficients. The use of $3j$ symbols allows these matrix elements to be deduced simply and directly. The method we use above, though less elegant, is more appropriate to our introductory approach where we are attempting to illustrate the assumptions involved at every stage rather than describe the most powerful techniques of evaluating these matrix elements.[8, 56]

and since $(e^{im'\phi})^* = e^{-im'\phi}$, the Φ integral is:

$$\int_0^{2\pi} e^{i(-m'+q+m)\phi}\, d\phi. \tag{2.49}$$

It is easily shown that this definite integral vanishes unless

$$(-m'+q+m) = 0,$$

when it equals 2π. All Φ parts of the matrix elements (2.46) therefore vanish or equal the same number, so we may write

$$\langle\psi_i|V|\psi_j\rangle \propto \langle\Theta_l^{m'}|\Theta_k^q|\Theta_l^m\rangle, \tag{2.50}$$

with the same proportionality constant for each matrix element of the set. The condition ensuring non-zero Φ integrals,

$$(-m'+q+m = 0),$$

means that non-zero Φ integrals exist for diagonal elements of the Y_4^0 term in the potential $(-m'+0+m = 0)$ and, since $q = 0$ or ± 4 from (2.44), that the only off-diagonal elements are between $d_{\pm 2}$ for $Y_4^{\pm 4}$.

The explicit forms of the harmonics, equivalent to (2.37), in V_{T_d} are

$$\left.\begin{aligned} Y_4^0 &= \left(\sqrt{\frac{1}{2\pi}}\right)\left(\sqrt{\frac{9}{128}}\right)(35\cos^4\theta - 30\cos^2\theta + 3), \\ Y_4^{\pm 4} &= \left(\sqrt{\frac{1}{2\pi}}\right)\left(\sqrt{\frac{315}{256}}\right)\sin^4\theta \, . \, e^{\pm 4i\phi}. \end{aligned}\right\} \tag{2.51}$$

Using (2.51) and (2.45), the matrix elements (2.50) may be written explicitly as follows:

$$\left.\begin{aligned} \langle d_0|V|d_0\rangle &= x.\frac{5}{8}\left(\sqrt{\frac{9}{128}}\right)\int_0^\pi (3\cos^2\theta - 1) \\ &\qquad \times (35\cos^4\theta - 30\cos^2\theta + 3)(3\cos^2\theta - 1)\sin\theta\, d\theta, \\ \langle d_{\pm 1}|V|d_{\pm 1}\rangle &= x.\frac{15}{4}\left(\sqrt{\frac{9}{128}}\right)\int_0^\pi (\cos\theta\sin\theta) \\ &\qquad \times (35\cos^4\theta - 30\cos^2\theta + 3)(\cos\theta\sin\theta)\sin\theta\, d\theta, \\ \langle d_{\pm 2}|V|d_{\pm 2}\rangle &= x.\frac{15}{16}\left(\sqrt{\frac{9}{128}}\right)\int_0^\pi (\sin^2\theta) \\ &\qquad \times (35\cos^4\theta - 30\cos^2\theta + 3)(\sin^2\theta)\sin\theta\, d\theta, \\ \langle d_{\pm 2}|V|d_{\mp 2}\rangle &= x.\frac{15}{16}\left(\sqrt{\frac{5}{14}}\right)\left(\sqrt{\frac{315}{256}}\right)\int_0^\pi (\sin^2\theta) \\ &\qquad \times (\sin^4\theta)(\sin^2\theta)\sin\theta\, d\theta, \end{aligned}\right\} \tag{2.52}$$

where x takes in some common parts of the normalizing coefficients, the radial integrals and the Φ integrals: the $\sqrt{\frac{5}{14}}$ in the final equation comes from the potential in (2.44). The Θ integrals in (2.52) can be worked out using the standard formula:

$$\int_0^\pi \sin^{2n+1}\theta \,.\, \cos^m\theta \,.\, d\theta = \int_0^\pi \cos^{2n+1}\theta \,.\, \sin^m\theta \,.\, d\theta = \frac{2^{n+1} \,.\, n!}{(m+1)(m+3)\ldots(m+2n+1)}. \tag{2.53}$$

Collecting factors in (2.52), we find the relations between the matrix elements are as follows:

$$\left.\begin{aligned} \langle d_0|eV|d_0\rangle &= 6y, \\ \langle d_{\pm1}|eV|d_{\pm1}\rangle &= -4y, \\ \langle d_{\pm2}|eV|d_{\pm2}\rangle &= y, \\ \langle d_{\pm2}|eV|d_{\mp2}\rangle &= 5y, \end{aligned}\right\} \tag{2.54}$$

where y, like x in (2.52), involves radial integrals, Φ integrals and sundry pure numbers. The secular equation for the d orbital set under V_{T_d} is therefore:

$$\begin{array}{c} \\ d_2 \\ d_1 \\ d_0 \\ d_{-1} \\ d_{-2} \end{array}\begin{array}{c} \begin{array}{ccccc} d_2 & d_1 & d_0 & d_{-1} & d_{-2} \end{array} \\ \left|\begin{array}{ccccc} y-E & 0 & 0 & 0 & 5y \\ 0 & -4y-E & 0 & 0 & 0 \\ 0 & 0 & 6y-E & 0 & 0 \\ 0 & 0 & 0 & -4y-E & 0 \\ 5y & 0 & 0 & 0 & y-E \end{array}\right| \end{array} = 0. \tag{2.55}$$

The roots E are: $E = +6y$, twice,

$E = -4y$, three times.

Back-substitution of these roots into the secular equations in the usual way gives:[a]

$$\left.\begin{array}{c} d_0 \\ \frac{1}{\sqrt{2}}(d_2 + d_{-2}) \end{array}\right\} \quad \text{at} \quad E = +6y,$$

$$\left.\begin{array}{c} d_{\pm1} \\ \frac{1}{\sqrt{2}}(d_2 - d_{-2}) \end{array}\right\} \quad \text{at} \quad E = -4y.$$

Pictorially we have the familiar result in figure 2.2. We note the emergence of the baricentre rule, i.e.

$$\sum_i^5 E_i = 0. \tag{2.56}$$

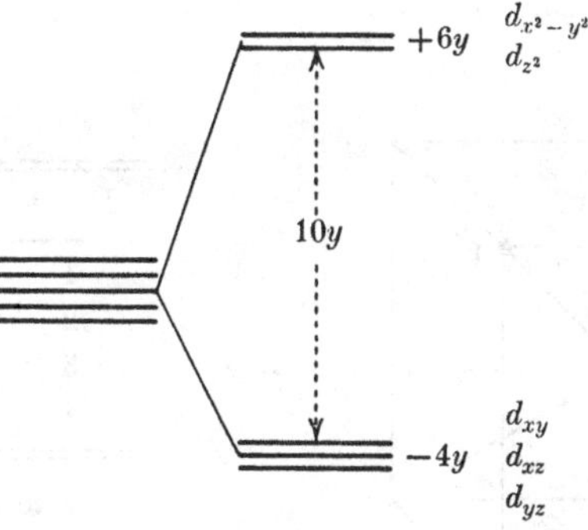

Figure 2.2.

Remember that neither the magnitude nor sign of y has been determined by the above procedures.

Taking stock and repeating an earlier point, we note that all that has been involved so far is the assumption that the ligands set up a potential such that the metal electrons are to be operated upon by a one-electron operator. The limitations set by the basis functions (d orbitals, for example) and the site symmetry of the complex molecules both determine, by group-theoretical methods, the form which an effective potential takes, e.g.

$$V_{T_d} = Y_4^0 + (\sqrt{\tfrac{5}{14}})\,(Y_4^4 + Y_4^{-4}).$$

The baricentre rule[b] follows from the neglect of Y_0^0 in V_{T_d}. The emergence of the splitting pattern in figure 2.2 is consistent with the simple group-theoretical result that a d orbital set spans the representations $e + t_2$ in T_d symmetry. So we see that if no more is assumed than that we are dealing with an electrostatic crystal-field of a particular symmetry then the writing down of a potential eventually tells us no more than a direct application of symmetry theory as in chapter 1. We can learn more of the sign and magnitude of y by specifying the nature of the crystal-field model further. This we now do.

THE POINT-CHARGE MODEL

The simplest electrostatic model we can describe is that which represents the ligands as point, negative charges. In figure 2.3 we depict a tetrahedral array of four point charges ze about a central metal ion. We do not specify anything of the radial nature of the metal wavefunctions.

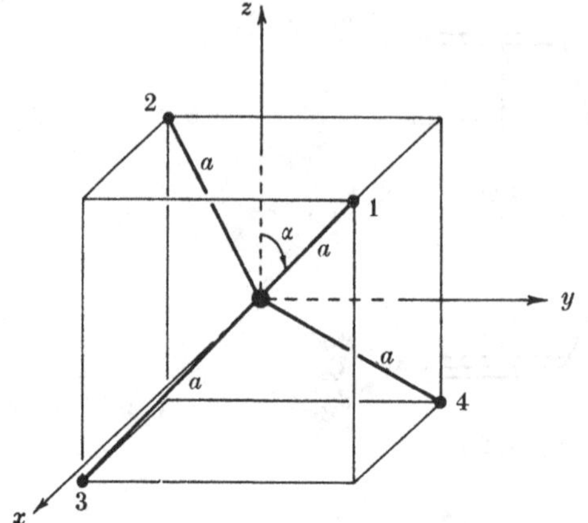

Figure 2.3. Axes and ligand positions in T_d symmetry.

TABLE 2.1

Point	r'	θ'	ϕ'
1	a	α	$\frac{1}{4}\pi$
2	a	α	$\frac{5}{4}\pi$
3	a	$\pi-\alpha$	$\frac{7}{4}\pi$
4	a	$\pi-\alpha$	$\frac{3}{4}\pi$

where $\alpha = \cos^{-1}\sqrt{\frac{1}{3}}$

The polar coordinates of the point charges are shown in table 2.1. For reasons which will become clear, when referring to the polar coordinates of the metal electrons we use (r, θ, ϕ) and for the ligands (r', θ', ϕ').

The potential experienced by a metal electron at some general point in space is the sum of terms which are proportional to the magnitudes of the point charges ze and inversely proportional to its distance from these point charges. If we write $\mathbf{r}_i$ as the vectorial distance of the ith ligand from the coordinate origin (the metal ion) and $\mathbf{r}_j$ as that of the general point in space, the potential may be written:

$$V = \sum_i^4 \frac{ze}{|\mathbf{r}_j - \mathbf{r}_i|}. \tag{2.57}$$

The expansion theorem allows us to express (2.57) as a linear combination of functions of the form $R_k(r) . Y_k^q(\theta, \phi)$ as in (2.17). This 'multipole expansion', as it is called, is a little difficult to prove and an exposition of this does not clarify the points we wish to make here. We shall discuss it a little more fully in the next chapter. We are content here to state the result[a] but in a form which looks different from that usually presented:†

† (*a*) The more conventional presentation [a, b] of this result is:

$$\sum_i \frac{1}{\mathbf{r}_{ij}} = \sum_i \sum_{l=0}^{\infty} \sum_{m=-l}^{l} \frac{4\pi}{2l+1} \cdot \frac{r^l}{a^{l+1}} \cdot Y_l^m(\theta_j, \phi_j) . Y_l^{m*}(\theta_i', \phi_i')$$

and we shall discuss it in roughly this form in the next chapter.

(*b*) Equation (2.58) and the one in (*a*) are appropriate for $r < r'$ ($r < a$): the slight untruth in these equations does not affect the present discussion and will be remedied in chapter 7.

$$\sum_i \frac{ze}{|\mathbf{r}_i-\mathbf{r}_j|} = \sum_{k=0}^{\infty}\sum_{q=-k}^{k}\frac{r^k}{r'^{k+1}}\left[\sum_i ze(\sqrt{2\pi})\left(\sqrt{\frac{2}{2k+1}}\right).\,Y_k^{q*}(\theta_i',\phi_i')\right]$$
$$\times(\sqrt{2\pi})\left(\sqrt{\frac{2}{2k+1}}\right).\,Y_k^q(\theta_j,\phi_j). \quad (2.58)$$

Equation (2.58) should be compared with (2.17). They are the same but for the fact that the adoption of the point-charge model (the multipole expansion) has now specified in some detail the forms of the expansion coefficients c_k^q and of the radial functions $R_k(r)$ in (2.17). To make them look more alike we can regroup terms in (2.58) as follows:

$$\sum_i \frac{ze}{|\mathbf{r}_i-\mathbf{r}_j|} = \sum_{k=0}^{\infty}\sum_{q=-k}^{k}\left[\frac{4\pi}{2k+1}\sum_i ze\,Y_k^{q*}(\theta_i',\phi_i')\right]\frac{r^k}{r'^{k+1}}.\,Y_k^q(\theta_j,\phi_j). \quad (2.59)$$

So that

$$\left.\begin{aligned} c_k^q &\equiv \frac{4\pi}{2k+1}ze\sum_i Y_k^{q*}(\theta_i',\phi_i'),\\ R_k(r) &\equiv \frac{r^k}{r'^{k+1}}.\end{aligned}\right\} \quad (2.60)$$

The presentation in (2.58) emphasizes the separation of parts associated with the points i and parts with the general point j. Again this will be made clearer in the next chapter.

However, the expansion in (2.58) has specified the 'unknowns' in (2.17) as in (2.60). They may be calculated explicitly by substituting in (2.60) the explicit values of (r',θ',ϕ') for the four ligands in figure 2.3. We could work them all out (up to Y_4^q if we are concerned with d orbitals) but we would find most vanish. Let us make use of the symmetry-based information we already have, concentrate on d orbital splittings and ignore Y_0^0 as before. We are left, therefore, with

$$V = \sum_i^4 \frac{ze}{|\mathbf{r}_i-\mathbf{r}_j|} = c_4^0.\,R_4(r).\,Y_4^0 + c_4^4.\,R_4(r).\,Y_4^4 + c_4^{-4}.\,R_4(r).\,Y_4^{-4}. \quad (2.61)$$

We require the explicit forms of these harmonics: they were given in (2.51). Using (2.51) and (2.60), we have:

$$c_4^0.\,R_4(r) = ze.\frac{4\pi}{9}.\frac{r^4}{a^5}.\left(\sqrt{\frac{1}{2\pi}}\right)\left(\sqrt{\frac{9}{128}}\right)\sum_i^4(35\cos^4\theta_i' - 30\cos^2\theta_i' + 3)$$
$$= ze.\frac{4\pi}{9}.\frac{r^4}{a^5}.\left(\sqrt{\frac{1}{2\pi}}\right)\left(\sqrt{\frac{9}{128}}\right).4.\left(-\frac{28}{9}\right). \quad (2.62)$$

Associated with $Y_4^{\pm 4}$ we have:

$$c_4^{\pm 4}.R_4(r) = ze.\frac{4\pi}{9}.\frac{r^4}{a^5}\left(\sqrt{\frac{1}{2\pi}}\right)\left(\sqrt{\frac{315}{256}}\right).\sum_i^4 \sin^4\theta_i'.e^{\pm 4i\phi_i'}$$

$$= ze.\frac{4\pi}{9}.\frac{r^4}{a^5}\left(\sqrt{\frac{1}{2\pi}}\right)\left(\sqrt{\frac{315}{256}}\right).[\tfrac{4}{9}(e^{\pm i\pi}+e^{\pm i\pi}+e^{\pm i\pi}+e^{\pm i\pi})]$$

$$= ze.\frac{4\pi}{9}.\frac{r^4}{a^5}\left(\sqrt{\frac{1}{2\pi}}\right)\left(\sqrt{\frac{315}{256}}\right).4.\frac{4}{9}. \tag{2.63}$$

Substituting (2.62) and (2.63) into (2.61) we get:

$$V_{T_d} = ze\left(-\frac{4}{9}\right)\left(\frac{7}{3}(\sqrt{\pi})\frac{r^4}{a^5}\right)\left[Y_4^0+\left(\sqrt{\frac{5}{14}}\right)(Y_4^4+Y_4^{-4})\right]. \tag{2.64}$$

The $\sqrt{\frac{5}{14}}$ coefficient of $Y_4^{\pm 4}$ relative to Y_4^0 emerges as previously established by symmetry.

We may now calculate the form of y in (2.54) from the point-charge model. Consider the matrix element:

$$M = \langle d_{+2}|eV|d_{-2}\rangle = \langle d_{-2}|eV|d_{+2}\rangle. \tag{2.65}$$

Substitution of (2.45) and (2.51) gives:

$$M = \langle d_2|-\tfrac{4}{9}.\tfrac{7}{3}(\sqrt{\pi})ze^2.\frac{r^4}{a^5}.\left(\sqrt{\frac{5}{14}}\right)Y_4^4|d_{-2}\rangle$$

$$= \langle R(d).\left(\sqrt{\frac{1}{2\pi}}\right)\left(\sqrt{\frac{15}{16}}\right).\sin^2\theta.e^{2i\phi}|-\tfrac{4}{9}.\tfrac{7}{3}(\sqrt{\pi})ze^2.\frac{r^4}{a^5}$$

$$\times\left(\sqrt{\frac{5}{14}}\right)\left(\sqrt{\frac{1}{2\pi}}\right)\left(\sqrt{\frac{315}{256}}\right).\sin^4\theta.e^{4i\phi}|R(d)$$

$$\times\left(\sqrt{\frac{1}{2\pi}}\right)\left(\sqrt{\frac{15}{16}}\right).\sin^2\theta.e^{-2i\phi}\rangle$$

$$= -\tfrac{4}{9}.ze^2\langle R(d)\left|\frac{r^4}{a^5}\right|R(d)\rangle.\frac{1}{2\pi}.\frac{15}{16}.\frac{7}{3}(\sqrt{\pi})\left(\sqrt{\frac{5}{14}}\right)\left(\sqrt{\frac{315}{256}}\right)$$

$$\times\langle\sin^2\theta|\sin^4\theta|\sin^2\theta\rangle.\langle e^{2i\phi}|e^{4i\phi}|e^{-2i\phi}\rangle.$$

Evaluating with aid from (2.53):

$$\langle\sin^2\theta|\sin^4\theta|\sin^2\theta\rangle = \int_0^\pi \sin^8\theta\sin\theta\,d\theta$$

$$= \frac{2^5.4!}{1.3.5.7.9} = \frac{256}{315},$$

$$\langle e^{2i\phi}|e^{4i\phi}|e^{-2i\phi}\rangle = \int_0^{2\pi} e^0 d\phi = 2\pi.$$

We write

$$\langle R(d)\left|\frac{r^4}{a^5}\right|R(d)\rangle \equiv \int_0^\infty R(d).\frac{r^4}{a^5}.R(d).r^2dr \equiv \overline{\left(\frac{r^4}{a^5}\right)}.$$

Then
$$M = -\tfrac{4}{9}.\tfrac{5}{6}.ze^2\overline{\left(\frac{r^4}{a^5}\right)}. \tag{2.66}$$

DEFINITIONS OF Δ_{oct}, Δ_{tet} AND $10Dq$

By comparison with (2.54),

$$y = -\tfrac{4}{9}.\tfrac{1}{6}ze^2\overline{\left(\frac{r^4}{a^5}\right)}.$$

As $\overline{(r^4/a^5)}$ may be regarded as related to a 'mean fourth-power radius' of the metal ion it is obviously a positive quantity and so y is seen to be negative. This means the splitting in figure 2.2 is inverted with the doublet lower than the triplet. We could define the splitting $e-t_2$ as $10Dq$ which gives $Dq = -\tfrac{4}{9}.\tfrac{1}{6}.ze^2\overline{(r^4/a^5)}$. However, we do not take this definition for the following reasons.

TABLE 2.2. *Octahedral point-charge coordinates*

	Cartesian			Polar		
Point	x	y	z	r'	θ'	ϕ'
1	a	0	0	a	$\frac{1}{2}\pi$	0
2	$-a$	0	0	a	$\frac{1}{2}\pi$	π
3	0	a	0	a	$\frac{1}{2}\pi$	$\frac{1}{2}\pi$
4	0	$-a$	0	a	$\frac{1}{2}\pi$	$\frac{3}{2}\pi$
5	0	0	a	a	0	0
6	0	0	$-a$	a	π	0

If we had considered an octahedral disposition of six point charges rather than the tetrahedral case above, the coordinates of these points would be as in table 2.2. The only difference between the potential for these (and other) cubic groups are the expansion coefficients c_k^q in (2.60). Thus:

$$\frac{c_4^0(T_d)}{c_4^0(O_h)} = \frac{\sum_i^4 Y_4^0(\theta_i',\phi_i')}{\sum_j^6 Y_4^0(\theta_j',\phi_j')} = \frac{\sum_i^4 (35\cos^4\theta_i' - 30\cos^2\theta_i' + 3)}{\sum_j^6 (35\cos^4\theta_j' - 30\cos^2\theta_j' + 3)}. \tag{2.67}$$

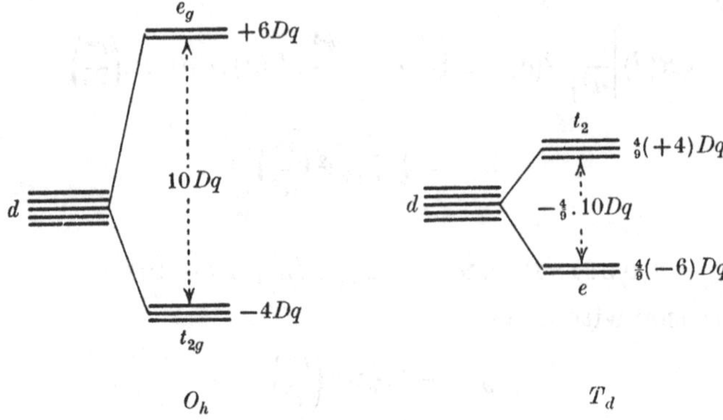

Figure 2.4. Crystal-field splittings for d orbitals in O_h and T_d symmetry.

The θ', ϕ' in the numerator of (2.67) are taken from table 2.1 and those in the denominator, from table 2.2. Hence:

$$\frac{c_4^0(T_d)}{c_4^0(O_h)} = \frac{4(-\frac{28}{9})}{(3+3+3+3+8+8)} = \frac{4(-\frac{28}{9})}{28} = -\frac{4}{9}. \qquad (2.68)$$

The same is true of the ratio $c_4^{\pm 4}(T_d)/c_4^{\pm 4}(O_h)$ as may be checked explicitly or by recalling the symmetry arguments used earlier. In other words, had we pursued the octahedral case, y in (2.54) would be $\frac{1}{6}.ze^2\overline{(r^4/a^5)}$. As the $(-\frac{4}{9})$ factor between octahedral and tetrahedral crystal fields comes from geometrical factors (the tables of polar coordinates), we define Dq to involve the energy separation between t_{2g} and e_g orbitals in the octahedron. Accordingly, we take

$$Dq \equiv \tfrac{1}{6}.ze^2\left(\frac{\overline{r^4}}{a^5}\right), \qquad (2.69)$$

so that $y(O_h) = Dq$ and $y(T_d) = -\frac{4}{9}Dq$, and there emerge the splitting patterns shown in figure 2.4. We shall refer to the orbital splittings in O_h and T_d symmetry as Δ_{oct} and Δ_{tet} and retain the term Dq to refer to the radial parameter (2.69).

SUMMARY AND OUTLOOK

Let us review what has been established so far. Assuming only a crystal-field potential, we determined the angular part of that potential from group theory. We could also have established the splitting of the d orbital manifold into e and t_2 sets by the usual means and hence required a single splitting factor which could be called, arbitrarily, Δ_{tet}. By restricting the form of the potential to that set up by point

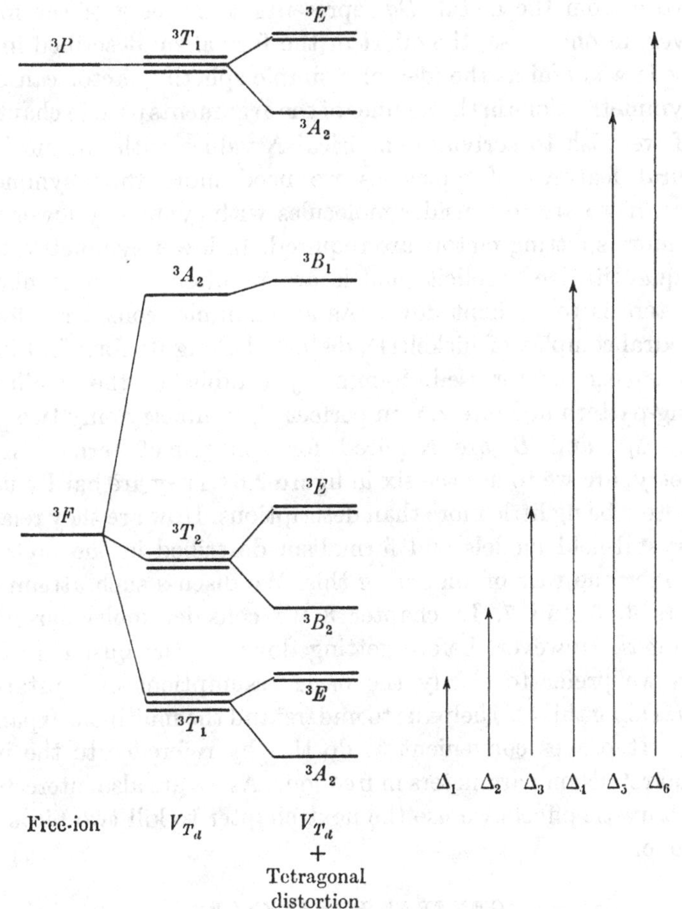

Figure 2.5. Splitting diagram for spin-triplet terms in tetragonally-distorted, tetrahedral nickel(II) systems.

charges we deduced the radial part of the potential and the form of $10Dq$ given in (2.69). A further restriction which defined the explicit form of the metal radial wavefunctions, i.e. $R(d)$, would finally establish $\overline{(r^4/a^5)}$ and all details of $10Dq$, which would then, of course, cease to be a parameter. Chapter 6 describes attempts to calculate $10Dq$ from first principles.

Thus we see something of the nature of Dq. With the minimum specification of the crystal field as a potential, Dq is a parameter expressing the nature of the metal–ligand interaction insofar as it affects spectral splittings. With the assumption of point-charges ze at distance a from the metal, Dq represents a property of the metal. However, in one sense, the effort of the formalism described in this chapter is wasteful as the idea of a simple splitting factor can come from symmetry alone in the manner of the arguments used in chapter 1. But, if we wish to correlate empirical Δ values with chemical and structural features of molecules we need more than symmetry. Further, if we are to consider molecules with symmetry lower than cubic, more splitting factors are required. In lower symmetry, then, more quantitative, explicit models are required if the number of parameters is to be kept down. As an example, consider a formal tetrahedral complex of nickel(II), distorted along its four-fold inversion axis – e.g. compressed. Symmetry establishes the qualitative splitting pattern in figure 2.5. In perfect T_d symmetry only two parameters – Δ_{tet} and B – are required for spin-triplet terms. In D_{2d} symmetry, are we to use the six in figure 2.5? They are hardly useful parameters being little more than descriptions. How are they related? The crystal-field models and formalism described in the preceding pages offer one way of answering this. We discuss such attempts in chapters 4, 5 and 7. In chapter 8 we consider molecular-orbital approaches. However, before getting down to the business of distortion we prefer to clarify the basic assumptions of crystal-field theory. Doing this also helps us to understand the multipole expansion of $1/r_{ij}$. It proves convenient to do this by reference to the inter-electron repulsion parameters in free-ions. As we are also interested in Nephelauxetic effects we use the next chapter to kill two birds with one stone.

GENERAL REFERENCES

(*a*) H. Eyring, J. Walter and G. E. Kimball, *Quantum Chemistry*, Wiley, New York, 1944.

(*b*) C. J. Ballhausen, *Introduction to Ligand Field Theory*, McGraw-Hill, New York, 1962.

(*c*) B. N. Figgis, *Introduction to Ligand Fields*, Interscience, New York, 1966.

(*d*) F. A. Cotton, *Chemical Applications of Group Theory*, Interscience, New York, 1963.

3
INTERELECTRON REPULSION PARAMETERS

MANY-ELECTRON SYSTEMS

Conventionally the description of the electronic properties of atoms or free-ions takes place in two main stages. The first concerns the basis states of a single electron attracted to a central nucleus. The second stage investigates the problems of ions with more than one electron. In both cases the angular properties of the electronic wavefunctions may be completely specified in terms, say, of spherical harmonics. Perhaps the earliest acquaintance with the idea of *s, p, d*, ... orbitals for those with chemists' training comes from a consideration of solutions to the hydrogen atom Schrödinger equation,

$$\nabla^2\psi + \frac{8\pi^2\mu}{h^2}(E-V)\psi = 0, \tag{3.1}$$

(where μ is the reduced mass of the electron). In an elementary treatment (see, for example, ref. 12), the full solution is not given but the idea of factorizing the equation into R, Θ and Φ parts is introduced as are the l and m_l quantum numbers. One of the notions which unfortunately comes with such a gentle introduction to quantum numbers is that the whole subject is inextricably associated with energies and the hydrogen atom itself. In somewhat more advanced texts (see, for example, ref. 23) the hydrogen atom angular wavefunctions in the guise of spherical harmonics are deduced directly from a consideration of the commutation relationships between the angular momentum operators L^2, L_z, L_x and L_y. After these approaches one might be left with the idea that it is angular momentum properties which determine the form of spherical harmonics: but, of course, we can abstract further, for the derivations referred to would follow from any relevant operators with the same commutation relationships. Infinitesimal rotation operators in spherical symmetry are just such operators.

We have made this little review to remind ourselves that the angular properties of single-electron wavefunctions, so-called *orbitals*, in free-ions are determined by group theory. The absolute charges and masses of the electrons and nucleus are only relevant for the radial parts of orbital functions. The same is true for many-electron systems. The *terms* which arise from a given electron configuration may be deduced simply by considering what representations in the spherical group are spanned by the many-electron product-wavefunctions of the configuration. As the ions are spherical, bases for these representations still take on the form of spherical harmonics: hence, we write capital S, L etc. instead of the lower-case s, l used for orbitals. This is the basis of the vector coupling method. We gave an example of it, in the previous chapter when deducing the 'vector triangle rule'.

Most treatments of many-electron systems distinguish between filled and open shells, often referred to as inner and valence shells, respectively. The distinction is possible by virtue of the 'central-field approximation'. Our treatment in the present chapter is relatively elementary and so a full definition and consequences of this approximation are not given; interested readers might consult references *a* and *b*. For our present purposes we merely state that one supposes that each electron moves in a spherical field due to the nucleus and all other electrons. We can accordingly describe the individual electron wavefunctions analogously to those for the hydrogen atom. The Hamiltonian for an n electron system may be written† as:

$$\mathscr{H}' = \sum_{\kappa=1}^{n} U(\kappa) + \sum_{\kappa<\lambda}^{n} V(\kappa, \lambda), \tag{3.2}$$

in which $U(\kappa)$ is a one-electron operator and $V(\kappa, \lambda)$ a two-electron operator. Specifically,

$$U(\kappa) = \frac{p^2(\kappa)}{2m} - \frac{Ze^2}{r(\kappa)} \tag{3.3}$$

describes the hydrogen-like Hamiltonian of kinetic energy plus nuclear attraction. It is a one-electron operator because it describes how a single electron (κ) moves in a potential field set up by an exterior source – the nucleus or nucleus plus electron core. On the other hand,

$$V(\kappa, \lambda) = \frac{e^2}{r(\kappa, \lambda)} \tag{3.4}$$

is a two-electron operator describing the mutual interaction of two of the n electrons (κ and λ) separated by a distance $r(\kappa, \lambda)$. Although the κth electron experiences a potential from an outside source – the λth electron – the fundamental indistinguishability of electrons requires us to use a mutual, two-electron, operator.

† In this chapter, electrons are generally labelled with Greek letters (except for n representing the total number of electrons) and wavefunctions by Roman letters as dummy indices and/or quantum numbers.

Matrix elements of U have the general form

$$\langle\psi_i|U|\psi_j\rangle = U_{ij}. \tag{3.5}$$

There are several important properties[a] of such matrix elements which derive from a formal treatment and these include the fact that only diagonal elements U_{ii} are non-zero in a free-ion configurational basis. Further, because

$$U_{ii} = \sum_{\kappa=1}^{n} \langle R_{nl}(\kappa)|U(\kappa)|R_{nl}(\kappa)\rangle, \tag{3.6}$$

all U_{ii} for ψs belonging to the same configuration, i.e. same n, l, are equal. This means that the effect of the U term in the Hamiltonian is to cause a uniform energy shift to all terms arising from a given configuration: that is, U is not responsible for any term splitting. Term splitting originates from the interelectron repulsion operator, V.

MATRIX ELEMENTS OF V

We therefore wish to examine matrix elements of the form,

$$\langle\psi_i|V|\psi_j\rangle = V_{ij} = \sum_{\kappa<\lambda}^{n} \langle\psi_i|V(\kappa,\lambda)|\psi_j\rangle. \tag{3.7}$$

The wavefunctions ψ are many-electron component term wavefunctions involving products like $\phi_1\phi_2\phi_3\ldots\phi_n$, where the ϕs are spin-orbitals. The fundamental indistinguishability of electrons demands that term functions be used which involve all permutations of the n electrons amongst the n one-electron spin-orbitals. The usual way of doing this is to write ψ_i as a Slater determinant:[23]

$$\psi_i = \frac{1}{\sqrt{(n!)}} \begin{vmatrix} \phi_1^i(1) & \phi_1^i(2) & \ldots & \phi_1^i(\kappa) & \ldots & \phi_1^i(n) \\ \phi_2^i(1) & \phi_2^i(2) & \ldots & \phi_2^i(\kappa) & \ldots & \phi_2^i(n) \\ \vdots & & & & & \\ \phi_n^i(1) & \phi_n^i(2) & \ldots & \phi_n^i(\kappa) & \ldots & \phi_n^i(n) \end{vmatrix} \tag{3.8}$$

or, in the usual shorthand:

$$\psi_i = |\phi_1^i(1)\phi_2^i(2) \quad \ldots \quad \phi_n^i(n)|. \tag{3.9}$$

The matrix element (3.7) is therefore written as:

$$V_{ij} = \langle|\phi_1^i(1)\phi_2^i(2)\ldots\phi_n^i(n)|\,.\,|V|\,.\,|\phi_1^j(1)\phi_2^j(2)\ldots\phi_n^j(n)|\rangle. \tag{3.10}$$

Now V operates on two electrons at a time so that (3.10) breaks down into the products:

$$\langle\phi_1^i(1)|\phi_1^j(1)\rangle \times \ldots \times \langle\phi_a^i(1)\phi_b^i(2)|V(1,2)|\phi_c^j(1)\phi_d^j(2)\rangle \times \ldots \times \langle\phi_n^i(1)|\phi_n^j(1)\rangle. \tag{3.11}$$

The matrix element V_{ij} involves the products (3.11) in which electrons 1 and 2 are placed in the differing orbitals ψ_i and ψ_j and so a sum over all electrons is not required. Because of the orthogonality of the single-electron functions ϕ, V_{ij} matrix elements are non-zero only if ψ_i and ψ_j differ by no more than two ϕ functions and so we need not sum over all orbital pairs in (3.11) either. If the central integral in (3.11) is non-zero, so also in general is that involving ϕ_a^i and ϕ_b^i interchanged. Such an interchange of orbitals† corresponds to an odd permutation of the Slater determinant (3.8) and so a minus sign appears in the general expression for the case where these ϕs have been interchanged. The preceding remarks are summarized in the expression:

$$\langle\psi_i|V|\psi_j\rangle = \pm[\langle\phi_a^i(1)\phi_b^i(2)|V(1,2)|\phi_c^j(1)\phi_d^j(2)\rangle - \langle\phi_b^i(1)\phi_a^i(2)|V(1,2)|\phi_c^j(1)\phi_d^j(2)\rangle]. \quad (3.12)$$

The $\pm$ signs outside the whole expression reflect the fact that an even or odd permutation may have been necessary to bring the Slater determinants ψ_i and ψ_j into the same standard order in which electrons 1 and 2 occupy the non-identical orbitals of the two determinants.

We get a slightly different expression if ψ_i and ψ_j differ in only one orbital. We can see this as follows. The integrals in (3.12) involve pairs of orbitals. If we regard ϕ_a^i as the 'odd man out' in ψ_i and ψ_j,

$$\left.\begin{aligned} \psi_i &= |\phi_1^i\phi_2^i \;\ldots\; \phi_a^i \;\ldots\; \phi_n^i|, \\ \psi_j &= |\phi_1^i\phi_2^i \;\ldots\; \phi_a^j \;\ldots\; \phi_n^i|, \end{aligned}\right\} \quad (3.13)$$

the ϕ_a^i may be 'paired' with $(n-1)$ other ϕ^is in these integrals. The expression for V_{ij} now involves a sum over all these possibilities:

$$\langle\psi_i|V|\psi_j\rangle = \pm\sum_t^{n-1}[\langle\phi_a^i(1)\phi_t^i(2)|V(1,2)|\phi_a^j(1)\phi_t^i(2)\rangle - \langle\phi_t^i(1)\phi_a^i(2)|V(1,2)|\phi_a^j(1)\phi_t^i(2)\rangle]. \quad (3.14)$$

Finally we have the diagonal case in which ψ_i and ψ_j are the same, say:

$$\psi_i = \psi_j = |\phi_1\phi_2 \;\ldots\; \phi_n|. \quad (3.15)$$

The same number of permutations is required to send each of ψ_i and ψ_j into a common order so that there is only a $+$ sign outside the whole expression this time. Since we may now consider any two ϕs at a time we have to sum over all possibilities which are distinct: to do this we keep $a < b$ in the sum over pairs of orbitals a and b, *viz.*

$$\langle\psi_i|V|\psi_j\rangle = \sum_{a<b}^{n}[\langle\phi_a(1)\phi_b(2)|V(1,2)|\phi_a(1)\phi_b(2)\rangle - \langle\phi_b(1)\phi_a(2)|V(1,2)|\phi_a(1)\phi_b(2)\rangle]. \quad (3.16)$$

† See ref.(*a*) for a discussion of the roles of electrons and orbitals.

The integrals in (3.16) have special names and symbols:

$$\left.\begin{aligned} \text{Coulomb integral } J(a,b) &\equiv \langle \phi_a \phi_b | V | \phi_a \phi_b \rangle, \\ \text{exchange integral } K(a,b) &\equiv \langle \phi_b \phi_a | V | \phi_a \phi_b \rangle. \end{aligned}\right\} \tag{3.17}$$

The reasons for this nomenclature are roughly as follows. If we factorize the ϕs into spatial and spin coordinates, f and τ_s,

$$\phi = f . \tau_s$$

and recall that

$$V(\kappa, \lambda) = \frac{e^2}{r_{\kappa\lambda}}$$

operates only on spatial coordinates, then

$$J(a,b) = \int \frac{e^2 |f_a(1)|^2 |f_b(2)|^2}{r_{12}} \, d\tau_1 d\tau_2, \tag{3.18}$$

which is the classical expression for the Coulomb interaction between two charge clouds of density $e|f_a|^2$ and $e|f_b|^2$. Accordingly, the integral $J(a,b)$ is called a Coulomb integral. There is no classical analogue for $K(a,b)$: because it involves the exchange of the two orbitals a and b, it is called the exchange integral. The exchange integral owes its existence to the basic principle of electrons being indistinguishable and is a purely quantum-mechanical feature.

In summary we may write for the diagonal element of V:

$$V_{ii} = \sum_{a<b}^{n} [J(a,b) - K(a,b)]. \tag{3.19}$$

EVALUATION OF V MATRIX ELEMENTS

Let us now consider how to evaluate an integral V_{ij} explicitly. We use the abbreviated notation:

$$V_{ij} = \langle \phi_a \phi_b | V | \phi_c \phi_d \rangle. \tag{3.20}$$

We proceed by *expanding the two-electron operator V as a series of products of one-electron operators*. We use[23] the expansion of $1/r_{12}$. In figure 3.1 we show the polar coordinates of two points in space representing the instantaneous positions of electrons 1 and 2. Simple trigonometry gives the distance between the points 1 and 2 as:

$$r_{12} = (r_1^2 + r_2^2 - 2r_1 r_2 \cos\gamma)^{\frac{1}{2}}, \tag{3.21}$$

and hence:

$$V(1,2) = \frac{e^2}{r_{12}} = \frac{e^2}{(r_1^2 + r_2^2 - 2r_1 r_2 \cos\gamma)^{\frac{1}{2}}}. \tag{3.22}$$

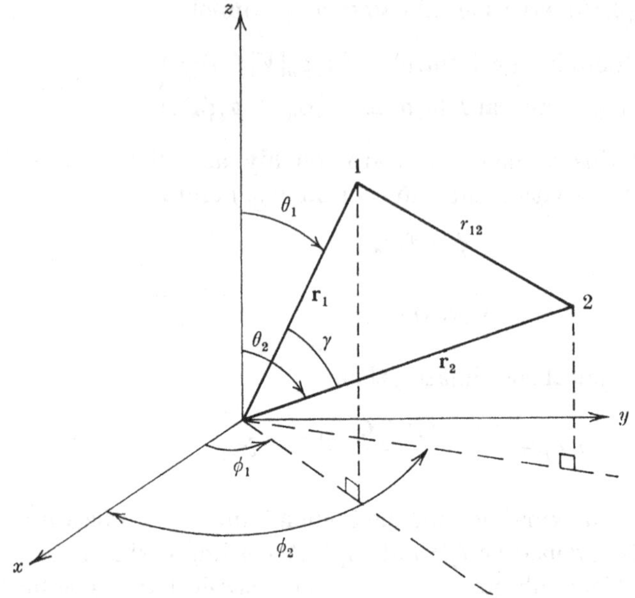

Figure 3.1. Polar coordinates for general points 1 and 2.

Defining $r_>$ and $r_<$ as the greater and lesser of r_1 and r_2, respectively we have:

$$V(1,2) = \frac{e^2}{r_>}\left(1+\frac{r_<^2}{r_>^2}-2\frac{r_<}{r_>}\cos\gamma\right)^{-\frac{1}{2}},$$

$$= \frac{e^2}{r_>}\sum_k^\infty \frac{r_<^k}{r_>^k} P_k(\cos\gamma). \tag{3.23}$$

This last step expressed the fact that $(1-2ax+x^2)^{-\frac{1}{2}}$ can be expanded as a linear combination of Legendre polynomials in the form $\sum_k^\infty x^k P_k(a)$. It is sufficient for our present purposes to view this expansion as another form of the expansion theorem discussed in the previous chapter. We now expand the Legendre polynomial $P_k(\cos\gamma)$ in terms of products of spherical harmonics using the spherical harmonic addition theorem.[2, 23] We do not reproduce the proof here but we note that the process is rather like the vector coupling procedure but in reverse. In effect we are expressing a coupled vector in terms of the products of uncoupled one-electron functions: i.e. akin to

$$Y_L^M \to \Sigma a Y_l^m(1) \,.\, Y_{l'}^{m'}(2). \tag{3.24}$$

This is relevant because a spherical harmonic Y_L^M may be written as a product of a Legendre polynomial (the Θ function) and a normalized Φ function ($Ne^{im\phi}$). The Legendre polynomial $P_k(\cos\gamma)$ is a function of the usual polar coordinates in figure 3.1, viz.

$$\cos\gamma = \cos\theta_1 . \cos\theta_2 + \sin\theta_1 . \sin\theta_2 . \cos(\phi_1 - \phi_2), \qquad (3.25)$$

and it can be shown[a, 23] that the products in (3.24) have the requirement that $l' = l$ and $m' = -m$ (or $m' = m^*$) and we write:

$$P_k(\cos\gamma) = \frac{4\pi}{2k+1} \sum_{q=-k}^{k} Y_k^q(1) . Y_k^{q*}(2). \qquad (3.26)$$

Perhaps the most important point in this expansion for our present purposes is the way V is expressed as a set of *products* of spherical harmonics. We referred to a 'reversed vector coupling approach' above. The 'normal' vector coupling may be illustrated by (2.30) of the previous chapter in that the products of two functions with angular momentum l and l' yields functions of 'total' angular momentum ranging $l+l'$ to $|l-l'|$. In the 'reverse' sense, we can see that a 'total' function with quantum numbers L and M can be arrived at by taking several appropriate products $|l, m\rangle . |l', m'\rangle$. For example, in the table of microstates for the d^2 configuration, there are three product functions which make up the state $|M_L = 0, M_S = 0\rangle$:

$$|0, 0\rangle = a(0^+, 0^-) + b(1^+, 1^-) + c(2^+, 2^-). \qquad (3.27)$$

Equations (3.24) and (3.26) express this sort of expansion.

The complete expansion of $1/r_{12}$ is then, from (3.23) and (3.26):

$$\frac{1}{r_{12}} = \sum_k^{\infty} \sum_{q=-k}^{+k} \frac{4\pi}{2k+1} . \frac{r_<^k}{r_>^{k+1}} . Y_k^q(1) . Y_k^{q*}(2), \qquad (3.28)$$

and we may now evaluate matrix elements of the two-electron operator V.

Explicitly, we wish to evaluate

$$\langle \phi_a \phi_b | V | \phi_c \phi_d \rangle = \left\langle \phi_a(1) \phi_b(2) \left| \frac{e^2}{r_{12}} \right| \phi_c(1) \phi_d(2) \right\rangle. \qquad (3.29)$$

The one-electron functions ϕ are written as products of radial, angular and spin parts, e.g.

$$\phi_a = R_{n_a l_a} . Y_{l_a}^{m_a} . \tau_{s_a}. \qquad (3.30)$$

Equation (3.29) may be written more fully as:

$$\langle \phi_a \phi_b | V | \phi_c \phi_d \rangle$$

$$= \left\langle \phi_a(1) \phi_b(2) \left| \sum_k^{\infty} \sum_{q=-k}^{k} \frac{4\pi}{2k+1} . \frac{r_<^k}{r_>^{k+1}} . Y_k^q(1)\, Y_k^{q*}(2) \right| \phi_c(1) \phi_d(2) \right\rangle. \qquad (3.31)$$

Notice that functions of electrons 1 and 2 appear in both bra and ket and these are individually operated upon by operators referring to electrons 1 and 2. Further, the operators do not affect the spin parts of the wavefunctions so that (3.31) vanishes unless the spin of 1 is the same in the bra as in the ket, and similarly for electron 2. Thus spin may be factored out using the delta functions $\delta(s_a s_c).\delta(s_b s_d)$. In passing, we can thus see why exchange integrals are non-zero only for pairs of electrons with parallel spins. Expanding (3.31) further gives:

$$\begin{aligned}\langle\phi_a\phi_b|V|\phi_c\phi_d\rangle &= \sum_{k=0}^{\infty}\langle R_{n_al_a}.R_{n_bl_b}\left|\frac{e^2r_<^k}{r_>^{k+1}}\right|R_{n_cl_c}.R_{n_dl_d}\rangle\\ &\quad\times\langle Y_{l_a}^{m_a}(1).Y_{l_b}^{m_b}(2)\left|\frac{4\pi}{2k+1}\sum_{q=-k}^{k}Y_k^q(1).Y_k^{q*}(2)\right|Y_{l_c}^{m_c}(1).Y_{l_d}^{m_d}(2)\rangle\\ &\quad\times\delta(s_a.s_c).\delta(s_b.s_d)\\ &= \sum_{k=0}^{\infty}R^k(abcd).A^k.\delta(s_as_c).\delta(s_bs_d),\end{aligned} \tag{3.32}$$

where A^k only involves the angular parts of the wavefunctions and operators. Because $Y_k^q(1)$ operates on electron 1 and $Y_k^{q*}(2)$ only on electron 2 we can write:

$$\begin{aligned}A^k &= \frac{4\pi}{2k+1}\sum_{q=-k}^{k}\langle Y_{l_a}^{m_a}(1).Y_{l_b}^{m_b}(2)|Y_k^q(1).Y_k^{q*}(2)|Y_{l_c}^{m_c}(1).Y_{l_d}^{m_d}(2)\rangle\\ &= \frac{4\pi}{2k+1}\sum_{q=-k}^{k}\langle Y_{l_a}^{m_a}|Y_k^q|Y_{l_c}^{m_c}\rangle\langle Y_{l_b}^{m_b}|Y_k^{q*}|Y_{l_d}^{m_d}\rangle.\end{aligned} \tag{3.33}$$

As discussed in the previous chapter, integrals $\langle Y_l^m|Y_k^q|Y_{l'}^{m'}\rangle$ can be expressed as products $\langle\Theta_l^m|\Theta_k^q|\Theta_{l'}^{m'}\rangle\langle\Phi_m|\Phi_q|\Phi_{m'}\rangle$. The Φ integral vanishes unless $-m+q+m' = 0$. Thus from (3.33) we may deduce the selection rule that

$$\begin{aligned}-m_a+q+m_c &= 0 \quad \text{for the first integral,}\\ -m_b+q^*+m_d &= 0 \quad \text{for the second integral.}\end{aligned}$$

But $q^* \equiv -q$, so that

$$m_a-m_c = q = m_d-m_b,$$

i.e.

$$m_a+m_b = m_c+m_d. \tag{3.34}$$

If this condition is satisfied, the Φ integral equals† $1/2\pi$. Hence,

$$A^k = \frac{2}{2k+1}\langle\Theta_{l_a}^{m_a}|\Theta_k^q|\Theta_{l_c}^{m_c}\rangle.\langle\Theta_{l_b}^{m_b}|\Theta_k^q|\Theta_{l_d}^{m_d}\rangle\delta(m_a+m_b, m_c+m_d), \tag{3.35}$$

† Remember that each Φ function has a normalizing coefficient $1/\sqrt{(2\pi)}$.

where the sum $\sum_{q=-k}^{k}$ in (3.33) reduces to one term in (3.35) because (3.34) is satisfied only by $q = m_a + m_b$ etc. We may define

$$c^k(lm, l'm') \equiv \left(\sqrt{\frac{2}{2k+1}}\right) \langle \Theta_l^m | \Theta_k^{m-m'} | \Theta_{l'}^{m'} \rangle, \qquad (3.36)$$

so that,

$$A^k = c^k(l_a m_a, l_c m_c) \,.\, c^k(l_b m_b, l_d m_d) \,.\, \delta(m_a + m_b, m_c + m_d). \qquad (3.37)$$

[Note also that $c^k(l'm', lm) = \left(\sqrt{\frac{2}{2k+1}}\right) \langle \Theta_{l'}^{m'} | \Theta_k^{m'-m} | \Theta_l^m \rangle$

$$= (-1)^{m-m'} c^k(lm, l'm'), \qquad (3.38)$$

because of the Condon–Shortley phase convention (see ref. 2, p. 19).]

Collecting definitions together we have:

$$\langle \phi_a \phi_b | V | \phi_c \phi_d \rangle = \sum_{k=0}^{\infty} R^k(abcd) \,.\, c^k(l_a m_a, l_c m_c) \,.\, c^k(l_b m_b, l_d m_d)$$

$$\times \delta(m_a + m_b, m_c + m_d) \,.\, \delta(s_a s_c) \,.\, \delta(s_b s_d). \qquad (3.39)$$

In practice we need not sum to infinity as suggested by (3.39) because, as we know from the previous chapter, the integrals $\langle \Theta_l^m | \Theta_k^q | \Theta_{l'}^{m'} \rangle$ are non-zero only if l, k and l' form a vector triangle. Thus, for d electron configurations, we need sum c^ks only for $k \leqslant 4$, etc. Values for the c^ks are tabulated in various textbooks,[a, b, 2] so that all that remains are the radial integrals R^k. We say more about these later, but for the moment we note that their evaluation requires a knowledge of the detailed radial parts of the wavefunctions which are not determined (like the c^ks) by symmetry. It is more usual practice to regard R^ks as parameters to be fitted to experiment: they are related to the so-called interelectron repulsion parameters. Before discussing such matters, however, we interrupt our review of the free-ion to see what we may now understand more clearly about the crystal-field formalism of the previous chapter.

THE CRYSTAL-FIELD FUNDAMENTAL CHARACTER[a]

We have just seen how the interelectron repulsion operator e^2/r_{12} may be expressed as the sum of products of two one-electron operators and also how the existence of a non-zero Coulomb integral may be associated with a similar exchange integral as a result of the fundamental indistinguishability of electrons. We used a similar expansion of $1/r_{12}$ when discussing the multipole expansion for crystal fields in chapter 2 [equation (2.58)]. There is a fundamental difference in these two uses of this expansion, however. The interelectronic repulsion matrix elements involve products of c^ks, that is of one c^k for each of the two electrons interacting with one another. Both integrals appear on an equal footing because electrons are indistinguishable: the arrangement $\phi_a(1).\phi_b(2)$ is as equally likely to occur as $\phi_a(2).\phi_b(1)$ and hence the need for Coulomb *and* exchange integrals. In the crystal-field usage one electron is taken to be a fixed negative charge representing some point in a rigid crystal, or molecular, lattice. In other words we have assumed one of the two 'electrons' appearing in the expansion is in a known position and is different in this respect at least from the other electron. We have assumed that the electrons of the metal and of the ligand are distinguishable and that those of the ligands may be viewed classically because they are fixed. There are now no 'product operators', for one half of these products are evaluated explicitly in terms of the coordinates of the fixed ligands. Similarly there are no exchange terms because the ligand 'electrons' are not regarded as quantum-mechanical particles. There is no question of antisymmetrizing between the quantum-mechanical metal electrons and the classical ligand charges. We are interested in the potential of the ligands sensed by the metal electrons but not the other way round. Although we have introduced these ideas by reference to the point-charge formalism, they are, in fact, fundamental to the crystal-field model. Crystal-field theory is concerned with one-electron operators – taken singly, not as products – and this is what distinguishes it from molecular-orbital theory where, initially at least, all electrons are considered under the quantum-mechanical umbrella. The idea of a crystal-field *potential*, that is something set up by an external 'classical' medium, is the fundamental concept of crystal-field theory. Options which specialize by describing the source of the potential as point-charges, point-dipoles or whatever imply no fundamental change in philosophy: they are of the same ilk.

INTERELECTRON REPULSION PARAMETERS

Our main reason for discussing interelectron repulsion effects at this stage in the book was to develop further the significance of the $1/r_{12}$ operator used in multipole expansions. As discussed in a previous section this concentrates attention on the angular integrals A^k. We shall now say a little about the radial integrals R^k, partly for completeness' sake in the present chapter and partly to lay a foundation for our discussion of the Nephelauxetic effect in chapter 9.

In (3.32) we defined the radial parts of the matrix elements of V_{ij} as:

$$R^k(abcd) \equiv \left\langle R_{n_a l_a} . R_{n_b l_b} \left| \frac{e^2 r_<^k}{r_>^{k+1}} \right| R_{n_c l_c} . R_{n_d l_d} \right\rangle. \tag{3.40}$$

We are usually concerned with interelectron repulsion effects within a configuration of equivalent electrons and we define the Condon–Shortley interelectron repulsion parameters:

$$F^k = R^k(abcd). \tag{3.41}$$

Note that for general diagonal elements V_{ii} we require:

$$\left.\begin{aligned} F^k(n_a l_a, n_b l_b) &\equiv R^k(n_a l_a, n_b l_b, n_a l_a, n_b l_b) \\ \text{and} \quad G^k(n_a l_a, n_b l_b) &\equiv R^k(n_b l_b, n_a l_a, n_a l_a, n_b l_b) \end{aligned}\right\} \tag{3.42}$$

corresponding to the Coulomb and exchange integrals, respectively. But for equivalent electrons $n_a = n_b$ and $l_a = l_b$, so that

$$F^k(n_a l_a, n_b l_b) = G^k(n_a l_a, n_b l_b). \tag{3.43}$$

Accordingly, we merely require the F^k as defined in (3.41).

We note also that for Coulomb and exchange integrals V_{ii}, (3.39) reduces to

$$\left.\begin{aligned} J(\phi_a \phi_b) &= \sum_{k=0}^{\infty} a^k(l_a m_a, l_b m_b) . F^k(n_a l_a, n_b l_b), \\ K(\phi_a \phi_b) &= \delta(s_a, s_b) \sum_{k=0}^{\infty} b^k(l_a m_a, l_b m_b) . G^k(n_a l_a, n_b l_b), \end{aligned}\right\} \tag{3.44}$$

where $\quad a^k(l_a m_a, l_b m_b) = c^k(l_a m_a, l_a m_a) . c^k(l_b m_b . l_b m_b),$

and $\quad b^k(l_a m_a, l_b m_b) = [c^k(l_a m_a, l_b m_b)]^2. \qquad (3.45)$

From these definitions, therefore, for *equivalent* electrons we get:

$$\left.\begin{aligned} J(\phi_a, \phi_b) &= \sum_{k=0}^{2l} a^k F^k, \\ K(\phi_a, \phi_b) &= \delta(s_a, s_b) \sum_{k=0}^{2l} b^k F^k, \end{aligned}\right\} \tag{3.46}$$

where sums are taken over even values of k only. This latter point means we require F^0, F^2, F^4, etc., but not F^1, F^3 etc.

Now we began this section stating that our interest had moved to the radial parameters only – so why have we re-introduced the angular parts c^k (and their derivatives a^k, b^k)? Well it happens that the angular Θ integrals, c^k, reduce to simple fractions. For example, for

d–d interelectron repulsions the c^2s are multiples of $\frac{1}{7}$ and the c^4s are multiples of $\frac{1}{21}$. Accordingly, it is conventional to include the fractions produced in the a^ks and b^ks of (3.46) with the purely radial functions F^k, Condon and Shortley [b] defining

$$\left.\begin{aligned} F_0 &= F^0, \\ F_2 &= \tfrac{1}{49}F^2, \\ F_4 &= \tfrac{1}{441}F^4, \end{aligned}\right\} \tag{3.47}$$

for d–d interactions, for example. (Note a^ks and b^ks involve the c^k denominators squared.) Remember also that the vector triangle rule means we require F_0, F_2 and F_4 only for d–d interelectron repulsions, as we would require F_0, F_2, F_4 and F_6 for f–f electron interactions.

TABLE 3.1. *Term energies for d^2*

Term	Condon–Shortley	Racah
^1S	$F_0+14F_2+126F_4$	$A+14B+7C$
^1G	$F_0+4F_2+F_4$	$A+4B+2C$
^3P	$F_0+7F_2-84F_4$	$A+7B$
^1D	$F_0-3F_2+36F_4$	$A-3B+2C$
^3F	$F_0-8F_2-9F_4$	$A-8B$

The standard texts cited at the end of this chapter describe how energies of the particular term arising from, say, a d^n configuration may be calculated in terms of the Condon–Shortley F_k parameters. The processes are straightforward but as our main interest here is in the nature of ligand-field parameters, we refer those interested to these standard texts. For illustration, we quote the results for the d^2 configuration in table 3.1. Like the spherically symmetric term V_0^0 in a crystal-field potential which shifts but does not split term energies, F_0 uniformly shifts all free-ion terms. Thus only the two parameters F_2 and F_4 are required to define d term splittings due to interelectron repulsions in a spherical field. The various coefficients in table 3.1 are determined partly by the angular Θ integrals, c^k, and partly by the occasional presence or absence of exchange as well as Coulomb integrals. For example, the energy of the ^1G term is

$$E(^1G) = J(2,1)-K(2,1), \tag{3.48}$$

and that of the ^1D is

$$E(^1D) = 2J(2,0)+J(1,1)-J(2,2)-J(2,1)+K(2,1). \tag{3.49}$$

Also shown in table 3.1 are the d^2 term energies expressed in terms of Racah parameters. These latter are linear combinations of the Condon–Shortley parameters chosen in such a way as to make the intervals between terms of maximum spin multiplicity functions of one parameter only. They are defined as follows:

$$\left.\begin{aligned} A &= F_0 - 49F_4, \\ B &= F_2 - 5F_4, \\ C &= 35F_4. \end{aligned}\right\} \tag{3.50}$$

The $^3F - {}^3P$ separation in table 3.1 is thus $15B$ in Racah parameters but $15F_2 - 75F_4$ in Condon–Shortley parameters. Racah parameters are usually favoured on the grounds of these conveniences they allow. There are occasions, however, when their use can disguise simple trends in the more 'basic' Condon–Shortley parameters, as we see in chapter 9.

REFINEMENTS OF THE THEORY

The theory of interelectron repulsions summarized so far in this chapter represents a first-order approximation. There are two main refinements of the theory concerned with configuration interaction on the one hand and orbital polarization or correlation on the other. Before discussing them it is well to remind ourselves that the perturbation we have been discussing is a purely electrostatic effect involving as it does matrix elements of the operator $1/r_{ij}$. The two-electron nature of this operator reflects the fact that we must deal with antisymmetrized wavefunctions, leading inevitably to both Coulomb and exchange integrals in appropriate cases. While Coulomb integrals are associated with energy destabilization and interelectron repulsion, the exchange integrals produce stabilization and interelectronic attraction. Self-consistent-field calculations of metal ion wavefunctions of the Hartree type give functions which are expanded relative to those of Hartree–Fock type: this is because the Hartree method neglects exchange integrals.

In $3d^n$ transition-metal ions, for example, the energy splittings of the various terms amounts to several tens of thousands of wavenumbers. If, as we have stated, the theory of these term energies is only a first-order approximation, then we need to be cautious as the crystal-field and spin–orbit perturbations we apply subsequently to these terms are only an order of magnitude smaller. However, the refinements we are about to describe do not alter the number of terms nor the nature of the ground term.

The first refinement we consider is concerned with configuration interaction. Throughout the theory discussed above we have only considered the effects of U_{ij} (showing an overall shift in energy for a configuration) and V_{ij} *within* a single configuration. However, *both* these one- and two-electron operators may have non-zero matrix elements between different configurations. Even though the zeroth-order energies of different configurations may be quite different, these inter-configuration matrix elements may be important. A particularly famous example of this (see ref. *a*, p. 93) concerns the ^3D and ^1D terms arising from an excited configuration $3s3d$ in magnesium.

The energies of these two terms under V calculated by the theory outlined above are:

$$\left.\begin{aligned} E(^3D) &= F_0 - G_2, \\ E(^1D) &= F_0 + G_2, \end{aligned}\right\} \tag{3.51}$$

where $G_2 = \frac{1}{5}G^2(3s, 3d)$. (Note the requirement of both F_k and G_k integrals for these non-equivalent electrons.) As G_2 is positive, there is a clear prediction that ^3D lies lower than ^1D. Experimentally, the reverse is true. There is no question of explaining this away by invoking an incorrect assignment, as similar discrepancies are found for ^3D and ^1D terms arising from all $3snd$ ($n = 3$ to 12!) configurations.

As described by Griffith (ref. *a*, p. 93), Bacher sought to explain the discrepancy by considering the influence of the configuration $3p^2$ on $3s3d$. The $3p^2$ configuration gives rise to ^1S, ^1D and ^3P terms. The suggestion was that the ^1D from the $3p^2$ interacts with the ^1D from $3s3d$. The energies of the terms arising from these configurations are as follows:

$$\left.\begin{aligned} E(3s3d) &= I(3s) + I(3d) + F_0(3s, 3d)\begin{cases} +G_2(3s, 3d) & {}^1_aD \\ -G_2(3s, 3d) & {}^3D \end{cases} \\ E(3p^2) &= 2I(3p) + F_0(3p, 3p)\begin{cases} +10F_2(3p, 3p) & {}^1S \\ +F_2(3p, 3p) & {}^1_bD \\ -5F_2(3p, 3p) & {}^3P \end{cases} \end{aligned}\right\} \tag{3.52}$$

in which $I(3s)$, etc. represent the one-electron matrix elements $\langle\psi_{3s}|U|\psi_{3s}\rangle$ etc. which, together with the F_0 terms determine the configuration energies before the splitting by V_{ij} into terms.

The terms of similar symmetry may interact, *viz.* ^1_aD and ^1_bD and their energies are found by diagonalization of a 2×2 secular determinant of the form:

$$\begin{array}{c|cc|} & {}^1_aD & {}^1_bD \\ \hline {}^1_aD & H_{aa} - E & H_{ab} \\ {}^1_bD & H_{ab} & H_{bb} - E \end{array} = 0. \tag{3.53}$$

In the present case, calculations and experiment [the latter for the diagonal term in (3.53)] give:

$$H_{aa} = 52033, \quad H_{bb} = 65040, \quad H_{ab} = 11044\,\text{cm}^{-1}. \qquad (3.54)$$

This relatively large off-diagonal term causes the ^1D from $3s3d$ to lie lower than the ^3D, the final values being:

$$\left.\begin{array}{llll} E(^1D) = 71355 & \text{and} & 45718\,\text{cm}^{-1} & \text{(calculated from (3.53) and (3.54))}, \\ E(^3D) = & & 47957\,\text{cm}^{-1} & \text{(expt)}, \\ E(^1D) = & & 46403\,\text{cm}^{-1} & \text{(expt)}, \\ E(^1D) = & & 52033\,\text{cm}^{-1} & \text{(calculated without configuration interaction)}. \end{array}\right\} \qquad (3.55)$$

Another important point to emerge from this example is that the effect of configuration interaction is not only to shift energies but also to mix configurations. To this extent l is not generally a good quantum number for a given term ^{2S+1}L. This has consequences for those indulging in calculations of orbital reduction factors, for example (see ref. 41 and also chapter 9).

The example summarized above illustrates the point that some discrepancies in the simple theory may be explained by incorporating other configurations into the calculation. Unfortunately, simple though this process is to visualize, it does not really offer a very useful way of refining the basic theory. For if we include one other configuration we should include others and it is not clear initially where to stop for 'sufficient' accuracy. As F^k values are not readily calculable from scratch, the normal procedure is to fit observed spectra by a least-squares method. This can be done reasonably well for the simple theory as there are more spectral lines than parameters. As the theory is extended to include more excited configurations, however, this rapidly ceases to be so. Thus, in general, we must be satisfied with the simple theory furnishing correct term designations and correct ground terms, and bear in mind that term energies are not exact nor l quantum numbers perfectly 'good'. These discrepancies may be taken up in various interelectronic repulsion and crystal-field parameters, which we are, in any case, primarily concerned to compare from compound to compound. It is fortunate that the apparent errors in the simple theory are less for the later, transition elements than for elements of the first short period: the reasons for this are not well understood. Many such errors that remain in the transition series may often be 'removed' by the polarization correction we now discuss.

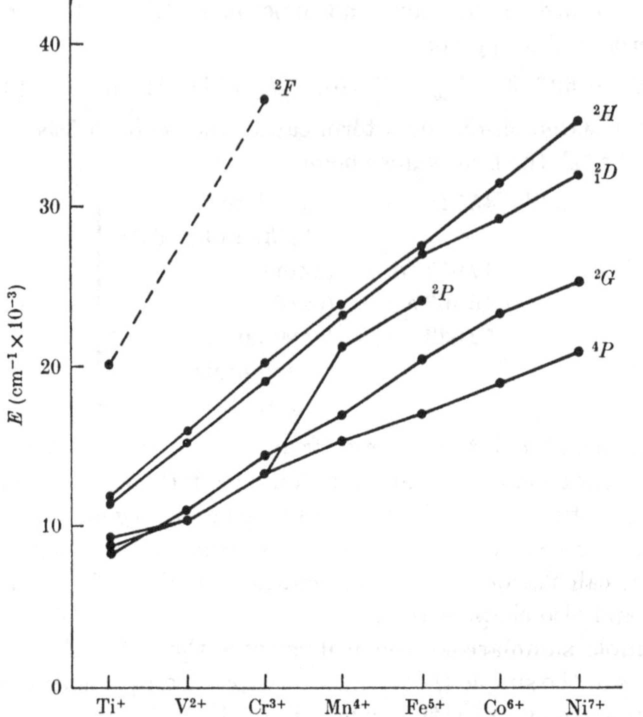

Figure 3.2. Energies of free-ion terms for iso-electronic (d^3) ions.[26]

THE TREES CORRECTION

The Trees correction (also called the 'polarization' correction for reasons which will become clear) was first proposed by Trees[109] who showed empirically that transition-metal free-ion spectra could be fitted to the simple interelectronic repulsion theory by inclusion of a term $\alpha L(L+1)$ in the diagonal term energies. Recent work of Ferguson and Wood[25, 26] illustrates the effect in complexes nicely. Figure 3.2 shows the experimental term energies for several $3d^3$ ions in the iso-electronic series Ti^+, V^{2+}, Cr^{3+}, Mn^{4+}, Fe^{5+}, Co^{6+}, Ni^{7+}. Assuming a constant ratio $F_4 = 0.07F_2$ in line with Watson's Hartree–Fock value, Ferguson and Wood calculate F_2 values for each term using relationships like those in table 3.1. They are plotted in figure 3.3, from which we observe that the F_2 values apparently increase with increasing L value of the terms. They applied a Trees correction of $\alpha L(L+1)$ to each term energy, that is; -10α, -6α, $+8\alpha$, $+18\alpha$ to the excitation energies of the ^4P, ^2D, ^2G and ^2H terms, respectively. (Note these corrections are

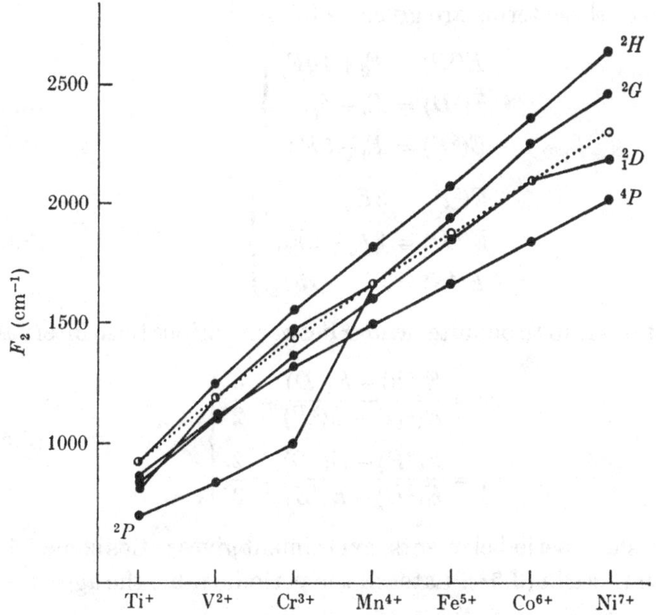

Figure 3.3. F_2 values for isoelectronic (d^3) ions.[26] Circles and dotted line show F_2 values after a Trees correction with $\alpha = 70\,\text{cm}^{-1}$.

applied to energy *differences*.) Taking $\alpha = 70\,\text{cm}^{-1}$ (the value previously determined for d^3 free-ions), the corrected F_2 values for each term concur very well, as shown by the circles in figure 3.3.

Similar treatments have been pursued for d^3, d^7; d^4, d^6; d^2, d^8 ions by Ferguson and Wood. In all cases, a Trees correction with $\alpha \sim 70\,\text{cm}^{-1}$ has allowed the energies of most of the terms arising from these configurations to be fitted with a single value of F_2; that is, to be fitted by the basic theory outlined earlier. The point of particular interest to these authors was that similar Trees corrections should be applied to data from complexes as a matter of course, before embarking upon an examination of Nephelauxetic effects (see chapter 9). There are a number of details and problems associated with the data summarized in figure 3.3 and those interested should consult the original papers.[25, 26] The point which the work illustrates, however, is the general efficiency of Trees' correction. We end this chapter with a few remarks concerning its physical origin.

Griffith[a] illustrates an argument of Racah concerned with the Trees' correction. The configurations p^2, p^4 each give rise to the terms ^1S, ^1D, ^3P; and p^3 to ^2P, ^2D and ^4S. According to the simple theory the energies of these terms are given as follows:

p^2 and p^4:
$$\left.\begin{aligned} E(^1S) &= F_0+10F_2,\\ E(^1D) &= F_0+F_2,\\ E(^3P) &= F_0-5F_2; \end{aligned}\right\} \tag{3.56}$$

p^3:
$$\left.\begin{aligned} E(^2P) &= 3F_0,\\ E(^2D) &= 3F_0-6F_2,\\ E(^4S) &= 3F_0-15F_2. \end{aligned}\right\} \tag{3.57}$$

These equations lead to definite ratios r between various term intervals, *viz.*
$$\left.\begin{aligned} r_2 = r_4 &= \frac{E(^1S)-E(^1D)}{E(^1D)-E(^3P)} = \frac{3}{2},\\ r_3 &= \frac{E(^2P)-E(^2D)}{E(^2D)-E(^4S)} = \frac{2}{3}. \end{aligned}\right\} \tag{3.58}$$

For the first short period elements, experiment gives ratios some 25 % smaller than those in (3.58). Later in the periodic table the agreement is better.

Now all term energy expressions simply involve the sum of two-electron functions because $1/r_{12}$ is a two-electron operator. One may then ask whether the observed deviations from the ratio rule for other p^n configurations can all be expressed solely in terms of the deviation for p^2. The details of the argument are not essential for our present purposes, but we note that the answer is affirmative. Racah has shown[88] that the term correction $\alpha L(L+1)$ is equivalent to each pair of electrons having an additional energy of the form $\alpha' l_1 . l_2$. For this reason the Trees correction is often referred to as due to orbit–orbit interaction.

The main electron–electron interactions described by the bulk of the theory in this chapter are due to the purely electrostatic interaction of the negatively charged quantum particles. Introductory texts describe these effects in terms of the 'vector coupling' model and frequently give the impression that their origin lies in the coupling of the magnetic moments of the electrons. Such impressions are unfortunate, as the effects are solely due to electrostatic interaction which is generally a much larger effect than magnetic interaction. However, electrons do possess magnetic moments and so magnetic coupling between electrons does occur in addition to the electrostatic interaction. The magnetic interactions give rise to those effects commonly described as spin–orbit, orbit–orbit and spin–spin coupling. Spin–orbit coupling is usually considered to be the most important of these effects: spin–spin coupling appears to give rise to extremely small perturbations indeed. To illustrate the order of magnitude of the various electron–electron interactions we note that for first-row transition elements, term separations due to the electrostatic $1/r_{12}$ operator are tens of thousands of wavenumbers; spin–orbit coupling causes splittings up to a thousand wavenumbers and often much less; spin–spin splittings in regular octahedral iron(III) compounds appears responsible for zero-field splittings in the range 0.001 to 0.05 cm^{-1}.

Racah's demonstration that

$$\alpha L(L+1)$$

is equivalent to $\alpha' l_1 . l_2$ suggested that the origin of the Trees correction lies in the magnetic, orbit–orbit interaction. However, Layzer[65] and also Yanagawa[119] have attempted to calculate the magnitude of such an effect in real systems. In both cases, their theories involved computation of the orbit–orbit interactions between electrons smeared out, as in the usual wavefunction description of electrons: they both calculated very small effects, roughly an order of magnitude less than observed by Trees. It was generally concluded on the basis of the results that the origin of the Trees correction is electrostatic: but a mechanism was not clear.

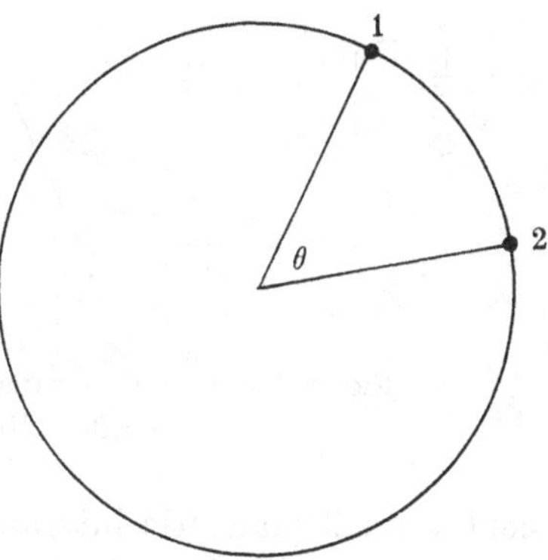

Figure 3.4.

However, Ufford and Callen[110] have resurrected the magnetic mechanism in an interesting way. They have shown that purely magnetic orbit–orbit coupling can give rise to effects of the required magnitude, provided that the positions and motions of the interacting electrons are *correlated*. They began by illustrating the effect in terms of a classical picture involving electrons in a Bohr orbit. In figure 3.4 we define an angle θ between the radius vectors of the two electrons. Ufford and Callen calculate the dependence of the magnetic orbit–orbit interaction as a function of θ, as in figure 3.5. Their simple classical model (subsequently replaced with a quantum-mechanical treatment) shows how the magnetic interaction changes sign as a function of the 'interelectron angle' θ. The qualitative conclusion they draw from this is that an uncorrelated wavefunction is apt to

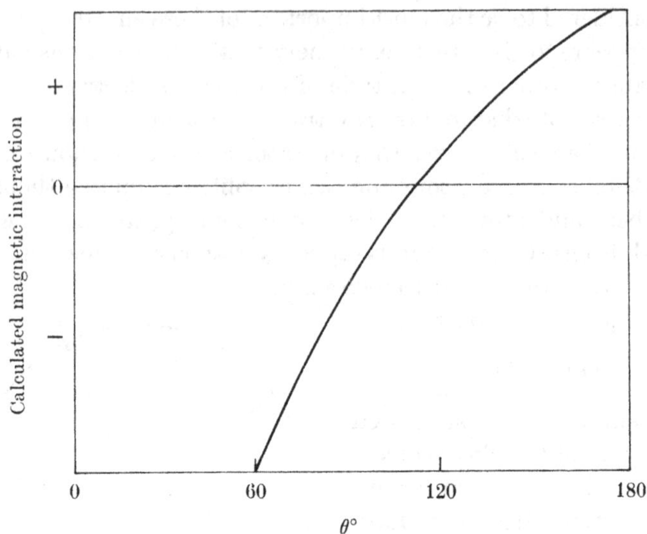

Figure 3.5. Calculated magnetic interaction as function of θ, after Ufford and Callen.[110]

yield a small magnetic interaction due to cancellation effects from large and small θ values. A correlated function, biassed against small θ, would lead to much larger values of the magnetic interaction. The authors also found that similar sign-change effects do not occur for other interactions – Coulomb, spin–orbit or spin–spin – and so conclude that these other interactions are much less sensitive to correlation of the electrons.

In summary, therefore, it appears that the Trees correction could have its origin in an orbit–orbit interaction resulting from the pair-wise coupling of electron magnetic moments, provided that such electrons are correlated in a manner corresponding to maximal rather than minimal interelectron separations. Ufford and Callen conclude that correlation would alter the electrostatic term values so that all interactions would have to be re-computed consistently in order to provide a comparison of theory and experiment. We have no idea how large such an alteration to the electrostatic energies would be. Griffith[a] makes a similar point in connection with the ^1S term of the $4d^2$ configuration of Zr. Fitting the basic theory for all ^3F, ^3P, ^1D, ^1G and ^1S terms gives:

$$B = 254, \quad C = 1423\ \text{cm}^{-1}.$$

If the ^1S term is omitted from the fitting procedure,

$$B = 254, \quad C = 1975\ \text{cm}^{-1}.$$

Griffith[a] observes that 'the bad fit of ^1S is probably due to the fact that the theoretical formulae predict ^1S to be high because the electrons are close together, while in the real eigenfunction the electrons adjust their positions relative to each other by including r_{12} explicitly in the eigenfunction'.

We are left, then, unable to say precisely whether current theories ascribe the Trees correction to electrostatic or magnetic effects. One way of looking at it would be to say that in correlated wavefunctions, the electrons electrostatically avoid each other (more than implied by a normal smeared function) so increasing the magnitude of their magnetic interactions toward the observed Trees correction. On the other hand, a recognition of electron correlation tending to minimize interelectron repulsions, implies a purely electrostatic mechanism. Though trite, it is probably best to say that both electrostatic and magnetic mechanisms play their part in Trees' correction.

GENERAL REFERENCES

(*a*) J. S. Griffith, *The Theory of Transition-Metal Ions*, Cambridge University Press, 1961.

(*b*) E. U. Condon and G. H. Shortley, *The Theory of Atomic Spectra*, Cambridge University Press, 1964.

4
RADIAL PARAMETERS FOR ANGULARLY-DISTORTED SYSTEMS

LOW SYMMETRY CRYSTAL-FIELD POTENTIALS DEFINED BY SYMMETRY

The utility of the multipole expansion or point-charge calculation in this subject is only realized in systems where symmetry leaves several unresolved factors. In chapter 2 we discussed the tetrahedron in which only one splitting factor remained to define the system after symmetry had been invoked. No particular model of point-charges, dipoles or of molecular orbitals aids the assignment of the *d–d* spectrum of a tetrahedral molecule any more than a mere definition of the splitting factor as Δ_{tet}, as in chapter 1. The model may aid conception and help to relate spectral properties with the nature of bonding etc., but that is another matter. On the other hand, if we consider molecules with symmetry low enough to produce several splitting factors, that is energy differences not mutually related, then to avoid the over-parametrized situation described at the end of chapter 2, we require a simple theory which correlates these factors. The point-charge model is such a theory. The low-symmetry situations we refer to may frequently be regarded as resulting from distortions imposed on octahedral or cubic molecules, the simplest category of distortion being axial. We illustrate all our points with axial distortions, although our conclusions should carry over to those systems suffering rhombic or lower-symmetry distortions also. The practical interpretation of results in the lower-symmetry conditions (that is, even the expression of direct observables in terms of the model parameters, let alone an understanding of the values of these parameters) is usually very difficult, if not impossible. Axial distortions are of two kinds, involving changes in bond lengths or in bond angles. For reasons we hope to make plain, we begin with this chapter on angular distortions. We could illustrate our arguments with trigonally-distorted octahedra or tetragonally-distorted tetrahedra, for example, as in figure 4.1. In each case all bond lengths are assumed equal and the distortion effected by a variable angle α subtended by any bond and the triad in the trigonally-distorted system or by any bond and the inversion tetrad in the tetragonally-distorted tetrahedron. In both

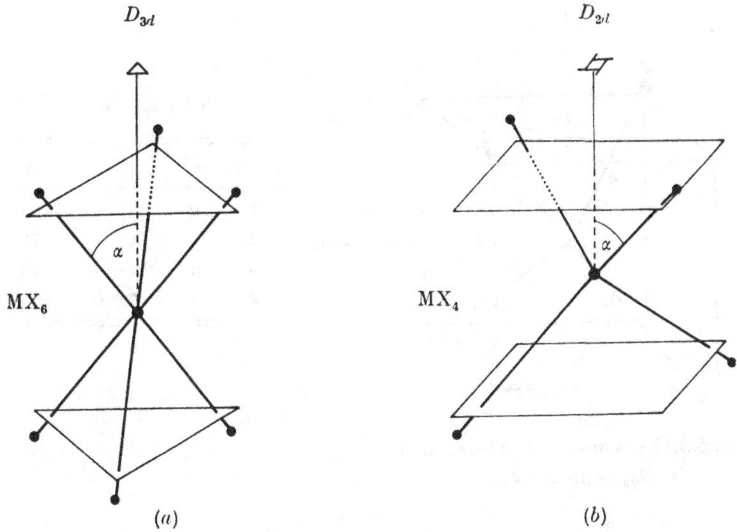

Figure 4.1. Trigonal octahedral (D_{3d}) and tetragonal tetrahedral (D_{2d}) distortions.

cases, exact cubic symmetry – O_h or T_d as appropriate – is achieved when α is exactly $\cos^{-1}\sqrt{\frac{1}{3}}(\sim 54.74°)$.

We shall illustrate the mathematics we need with the tetragonal case in figure 4.1(*b*). The trigonal case is fascinating, and we have things to say about it later, but the problem of defining axes would inevitably lead to a lengthy 'aside' demonstrating the equivalence of various authors' formulae without illuminating our main thesis. Accordingly, we discuss tetragonally-distorted tetrahedra and this is why we chose to illustrate chapter 2 with the tetrahedron rather than the octahedron.

The symmetry of the molecule in figure 4.1(*b*) belongs to the D_{2d} point-group. We investigate a point-charge crystal-field model of this symmetry using the point-coordinates shown in figure 4.2 and table 4.1. (Compare with figure 2.3 and table 2.1.)

As before, we begin by defining the potential in this symmetry. We illustrate the assertions of the first paragraph of this chapter by first seeking to define the potential from symmetry considerations alone. Once more, we consider matrix elements between d orbitals only. Hence we require even harmonics only in V, of order $\leqslant 4$. Ignoring the ever-present spherical term, we first study the second order term Y_2^m. Using (2.32), we find the characters of this (reducible) representation

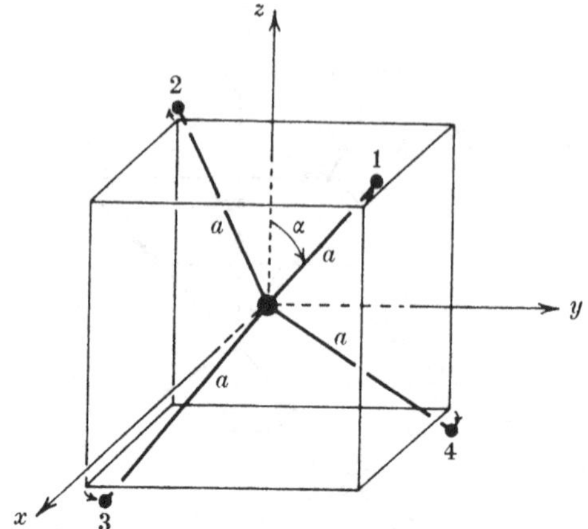

Figure 4.2. Axes and ligand positions in D_{2d} symmetry.

TABLE 4.1

Point	r'	θ'	ϕ'
1	a	α	$\frac{1}{4}\pi$
2	a	α	$\frac{5}{4}\pi$
3	a	$\pi-\alpha$	$\frac{7}{4}\pi$
4	a	$\pi-\alpha$	$\frac{3}{4}\pi$

under the elements of this group (cf. 2.33):

$$\begin{array}{l|ccccc} D_{2d} & E & 2S_4 & C_2 & 2C_2' & 2\sigma_d \\ \hline \chi(l=2) & 5 & -1 & 1 & 1 & 1 \end{array}. \qquad (4.1)$$

The irreducible components[17] of Y_2^m in D_{2d} are $A_1+B_1+B_2+E$ and so one combination of the Y_2^m is contained in the potential. The transformation properties of Y_2^m under $\hat{S}_4$ are as under $\hat{C}_4$, so by comparison with (2.35), we have:

$$\hat{C}_4\begin{bmatrix} Y_2^2 \\ Y_2^1 \\ Y_2^0 \\ Y_2^{-1} \\ Y_2^{-2} \end{bmatrix} = \begin{bmatrix} -Y_2^2 \\ iY_2^1 \\ Y_2^0 \\ -iY_2^{-1} \\ -Y_2^{-2} \end{bmatrix}. \qquad (4.2)$$

The trace under $\hat{C}_4$ must be unity so that only Y_2^0 appears in the potential. Turning to the fourth-order terms, we have:

$$\begin{array}{l|ccccc} D_{2d} & E & 2S_4 & C_2 & 2C_2' & 2\sigma_d \\ \hline \chi(l=4) & 9 & 1 & 1 & 1 & 1 \end{array} \qquad (4.3)$$

reducing to $2A_1+A_2+B_2+B_1+2E$, as is best checked using correlation tables. Thus *two* combinations of the Y_4^m transform as A_1 and therefore contribute to V. Note here, that the equivalent step in T_d symmetry forecast only *one* combination transforming as A_1. As S_4 symmetry elements are common to T_d and D_{2d} point groups, the next step is identical to that followed for T_d [equations (2.35), (2.36)], leading immediately to the fourth-order potential

$$V_4 = aY_4^4 + bY_4^0 + cY_4^{-4}. \tag{4.4}$$

Similarly, C_2 axes are common to T_d and D_{2d} symmetry so that the next step is identical to the T_d case also, giving by comparison with (2.37)–(2.39):

$$V_4 = bY_4^0 + c(Y_4^4 + Y_4^{-4}). \tag{4.5}$$

In the case of the tetrahedron, transformation under $\hat{C}_3$ fixes the ratio c/b as $\sqrt{(\frac{5}{14})}$. This had to happen because, as (2.34) showed, only one A_1 combination of the Y_4^m exists and so it is defined completely. But in the present case of symmetry D_{2d}, there are *two* A_1 representations and hence the ratio c/b cannot be fixed. This may be checked explicitly by noting that the C_3 axis in T_d vanishes in D_{2d} symmetry and there exists no other symmetry operation which may mix Y_4^0 and $Y_4^{\pm 4}$ as does $\hat{C}_3$ in T_d. The total potential, then, is:

$$V_{D_{2d}} = dY_2^0 + bY_4^0 + c(Y_4^4 + Y_4^{-4}). \tag{4.6}$$

No relationships between d and b or d and c are defined by symmetry as there are no symmetry operations possible apart from rotations and inversions (implied or actual) in any group: thus there are no operations which can transform one order of harmonics into another, e.g. Y_4^m into Y_2^m. We see, therefore, that D_{2d} symmetry defines a potential involving three unknowns as distinct coefficients or parameters.

LOW SYMMETRY AND THE POINT-CHARGE MODEL

We may relate these coefficients *via* the point-charge model. Actually, this only involves recasting d, b and c of (4.6) in terms of three other parameters. This would be an entirely pointless procedure if it were not possible to relate parameter values from one molecule to another and from one symmetry to another. It is with this aim in view that we proceed now with the point-charge calculation. Using the expansion of $1/r_{12}$:

$$\sum_i \frac{1}{r_{12}} = \sum_i \sum_{l=0}^{\infty} \sum_{m=-l}^{l} \frac{4\pi}{2l+1} \cdot \frac{r^l}{a^{l+1}} \cdot Y_l^m(\theta_2, \phi_2) \,.\, Y_l^{m*}(\theta_1, \phi_1), \tag{4.7}$$

recast as:

$$\sum_i \frac{1}{r_{12}} = \sum_{k=0}^{\infty} \sum_{q=-k}^{k} \left[\frac{4\pi}{2k+1} \sum_i Y_k^{q*}(\theta_1, \phi_1)\right] Y_k^q(\theta_2, \phi_2). \tag{4.8}$$

Using nomenclature established in chapter 2, this and table 4.1 give:

$$eV_{D_{2d}} = \frac{2(\sqrt{2\pi})}{5}\left(\sqrt{\frac{5}{8}}\right)4(3\cos^2\alpha-1).ze.\frac{r^2}{a^3}.\,Y_2^0$$
$$+\frac{2(\sqrt{2\pi})}{9}\left(\sqrt{\frac{9}{128}}\right)4(35\cos^4\alpha-30\cos^2\alpha+3).ze.\frac{r^4}{a^5}.\,Y_4^0$$
$$-\frac{2(\sqrt{2\pi})}{9}\left(\sqrt{\frac{315}{256}}\right)4\sin^4\alpha.ze\frac{r^4}{a^5}(Y_4^4+Y_4^{-4}). \quad (4.9)$$

Note that the last two terms in (4.9) are the same as those in (2.62) and (2.63) for T_d symmetry before the substitution of $\alpha = \cos^{-1}\sqrt{\frac{1}{3}}$. Thus, the ratio c/b in (4.5) and (4.6) depends on α which is not defined by symmetry. Similar remarks apply to the ratios d/b or d/c. *Two* things happen to the potential as a tetrahedron is tetragonally distorted: (1) the second-order term Y_2^0 is introduced and (2) the relative contributions of the fourth-order terms Y_4^0 and $Y_4^{\pm 4}$ alter. From (4.9) we notice that the radial parts (R) associated with Y_4^0 and $Y_4^{\pm 4}$ are both r^4/a^5 as emphasized in chapter 2, but associated with Y_2^0 there is a different radial function $-r^2/a^3$.

MATRIX ELEMENTS: THE DEFINITION OF Cp

We may now calculate the complete set of single-electron matrix elements $\langle m_l|V_{D_{2d}}|m_l'\rangle$ in a manner analogous to that in chapter 2 for V_{T_d}. Again the process involves products of three different types of integral with ϕ, θ and r arguments. The treatment of the Θ and Φ integrals is virtually identical with that used for T_d symmetry. The only new factor involved is that in evaluating integrals of the type $\langle m_l|V_2|m_l'\rangle$, i.e. $\langle m_l|(r^2/a^3)Y_2^0|m_l'\rangle$, radial integrals of the form

$$\left\langle R(d)\left|\frac{r^2}{a^3}\right|R(d)\right\rangle \quad (4.10)$$

occur, as opposed to

$$\left\langle R(d)\left|\frac{r^4}{a^5}\right|R(d)\right\rangle \quad (4.11)$$

for the fourth-order terms in V. We do not evaluate (4.10) explicitly, any more than (4.11) but analogously to the definition:

$$Dq = \tfrac{1}{6}ze^2\overline{\left(\frac{r^4}{a^5}\right)} = \tfrac{1}{6}ze^2\left\langle R(d)\left|\frac{r^4}{a^5}\right|R(d)\right\rangle,$$

we have:

$$Cp = \tfrac{2}{7}ze^2\overline{\left(\frac{r^2}{a^3}\right)} = \tfrac{2}{7}ze^2\left\langle R(d)\left|\frac{r^2}{a^3}\right|R(d)\right\rangle. \quad (4.12)$$

(The factor $\frac{2}{7}$ is introduced for consistency with an early paper on this subject.[44]) The matrix elements $\langle m_l|V_{D_{2d}}|m_l'\rangle$ are then given as follows:

$$\left.\begin{aligned}
\langle 0|V|0\rangle &= \tfrac{6}{7}Dq(35\cos^4\alpha-30\cos^2\alpha+3)\\
&\quad +2Cp(3\cos^2\alpha-1),\\
\langle \pm1|V|\pm1\rangle &= -\tfrac{4}{7}Dq(35\cos^4\alpha-30\cos^2\alpha+3)\\
&\quad +Cp(3\cos^2\alpha-1),\\
\langle \pm2|V|\pm2\rangle &= \tfrac{1}{7}Dq(35\cos^4\alpha-30\cos^2\alpha+3)\\
&\quad -2Cp(3\cos^2\alpha-1),\\
\langle \pm2|V|\mp2\rangle &= -5Dq\sin^4\alpha.
\end{aligned}\right\} \qquad (4.13)$$

DISTORTION PARAMETERS AND RADIAL PARAMETERS

Thus we see that instead of a potential parameterized by coefficients d, b, c as in (4.6), we have matrix elements (and hence, by implication, eigenvalues and eigenvectors) parameterized by α, Dq and Cp. Distortion of the tetrahedron is described by the angle α [or, actually, by $(\cos^{-1}\sqrt{\tfrac{1}{3}}-\alpha)$], while Dq and Cp are fourth- and second-order radial parameters defined by bond lengths and the exact nature of the metal radial wavefunctions. Insofar as bond lengths and metal orbital radial characteristics are substantially unaffected by (small) angular distortions from T_d symmetry, we see that the distortion is a purely angular property (α) and hence separable from the radial properties which depend only on the nature of the metal, ligand and bond length. This separability of a distortion parameter from the other, perhaps more fundamental, characteristics of radial parameters means that we should be able to examine a whole range of given MX_4 species in different crystal lattices and hopefully demonstrate Dq and Cp as common factors. More important still is that different MX_4 species may be compared once the distortion factors have been factored out. (In parenthesis and anticipating later discussions, the same is true of the angular overlap model, so that the foregoing is not to be seen as any specific justification for a point-charge model *per se*.) Another point at this stage is to note the similarity with results for a trigonally (i.e. angularly) distorted octahedron. The potential for D_{3d} symmetry, corresponding to the geometry in figure 4.1(a), may be written:†

$$\begin{aligned}
eV_{D_{3d}} &= \frac{4\pi}{5}\left(\sqrt{\frac{5}{8}}\right)\left(\sqrt{\frac{1}{2\pi}}\right)6(3\cos^2\alpha-1).ze\frac{r^2}{a^3}.Y_2^0\\
&\quad+\frac{4\pi}{9}\left(\sqrt{\frac{9}{128}}\right)\left(\sqrt{\frac{1}{2\pi}}\right)6(35\cos^4\alpha-30\cos^2\alpha+3)ze.\frac{r^4}{a^5}.Y_4^0\\
&\quad+\frac{4\pi}{9}\left(\sqrt{\frac{35}{16}}\right)\left(\sqrt{\frac{9}{4\pi}}\right)6\sin^3\alpha\cos\alpha.ze.\frac{r^4}{a^5}(Y_4^3-Y_4^{-3}),
\end{aligned} \qquad (4.14)$$

† But see ref. 36 for definition of axes.

and matrix elements $\langle m_l|V_{D_{3d}}|m_l'\rangle$ expressed as:

$$\left.\begin{aligned}\langle 0|V|0\rangle &= \tfrac{9}{7}Dq(35\cos^4\alpha-30\cos^2\alpha+3)\\ &\quad +3Cp(3\cos^2\alpha-1),\\ \langle\pm1|V|\pm1\rangle &= -\tfrac{6}{7}Dq(35\cos^4\alpha-30\cos^2\alpha+3)\\ &\quad +\tfrac{3}{2}Cp(3\cos^2\alpha-1),\\ \langle\pm2|V|\pm2\rangle &= \tfrac{3}{14}Dq(35\cos^4\alpha-30\cos^2\alpha+3)\\ &\quad -3Cp(3\cos^2\alpha-1),\\ \langle\mp2|V|\pm1\rangle &= \pm15Dq\sin^3\alpha\cos\alpha.\end{aligned}\right\} \tag{4.15}$$

Earlier remarks about the separability of angular and radial parameters apply for this geometry also, so permitting comparisons between trigonally-distorted octahedral and tetragonally-distorted tetrahedral geometries. Such comparisons would have been most obscure if potentials had merely been expressed in terms of the symmetry-defined parameters of (4.6), for example. As we shall see in the next chapter, such separability of distortion and radial parameters is not possible for non-angular distortions (e.g. tetragonal octahedral, D_{4h}) which is why we have introduced Cp before the parameters Ds and Dt.

Taking stock, we observe the introduction of second-order terms into the potential defined by axially-distorted octahedra or tetrahedra which lead, in the cases of angular distortions, to the parameterization of matrix elements and observables in terms of Cp, Dq and α. The distortion angle α describes a geometric feature which may have little chemical interest, perhaps resulting from peculiarities of crystal packing, which is separable in principle from the radial parameters Cp and Dq describing aspects of the ligand–metal relationships. We emphasize that both Cp and Dq are taken as *independent* parameters of the system despite a deceptively appealing relationship of the type:

$$Cp:Dq \propto \frac{r^2}{a^3} : \frac{r^4}{a^5} \quad ? \tag{4.16}$$

i.e. in which the bars over r^k/a^{k+1} have been omitted implying separability of r and a. We shall discuss the relationship of Cp to Dq in a later chapter from a theoretical point of view. Meanwhile we shall attempt to establish values of the ratio Cp/Dq determined empirically from spectral and magnetic studies. The remainder of this chapter is devoted to this and to illustrating the sensitivity of experimental observables to Cp.

It is appropriate, however, to first emphasize one aspect of our definitions, particularly of Dq. An alternative way of expressing (4.14) would be

$$V_{D_{3d}} = aY_2^0 + bY_4^0 + cV_{\text{oct}},$$

so that as distortion from O_h symmetry vanishes, so also do a and b. Matrix elements of the V_{oct} part would then involve, in a sense, the same definition of $10Dq$ as in octahedral symmetry (i.e. the energy separation of e_g and t_{2g} orbitals). The definition of Dq we give in (4.12), while identical with one tenth of the $e_g - t_{2g}$ energy difference in O_h symmetry, is different in D_{3d} symmetry if we express the potential as above. However, as t_{2g} and e_g are then no longer appropriate symmetry labels for the orbitals, there seems no compelling reason to maintain the O_h definition. By defining Dq as in (4.12) we hope to express a constant radial integral as independent of the degree of distortion, which is normally assumed to be small. A fundamental assumption here is that variations in α do not distort the metal orbitals. If this assumption is valid, the point-charge model (and point dipole and angular overlap models – see later) allows the coefficients a, b, c in the potential to be expressed as functions of changing α and constant Dq and Cp. If the assumption is not valid we might just as well express spectral and magnetic results in terms of the basic coefficients a, b and c. But, as we wish to correlate such values with those from other systems, without some basis such as the point-charge model, we might just as well merely record spectral transition energies direct.

Our definition of $10Dq$, which we use consistently throughout this book, is identical in practical terms with the $e_g - t_{2g}$ energy splitting, in octahedral symmetry. In other symmetries, however, it is different. Thus while we can equate $10Dq$ with Δ_{oct} in O_h symmetry, it does not equate with Δ_{tet} in T_d symmetry. There will be those who would prefer the quantity not to be called Dq: we consider that the possible confusion caused by the term Dq as we use it will be less than that caused by inventing a new symbol. The expression of $10Dq$ as the $t_{2g} - e_g$ splitting in the octahedral situation (where in any case the definitions agree) is of especial significance when comparing crystal-field and molecular-orbital calculations as we do at the end of chapter 6: but the problems of definitions raised there are not affected by the present discussion.

MAGNETIC PROPERTIES OF TRIGONALLY-DISTORTED OCTAHEDRA

We review here some studies of the magnetic susceptibilities and anisotropies of trigonally-distorted octahedral iron(II) and cobalt(II) molecules. The work[36] on iron(II) compounds grew from the observation of markedly dissimilar room-temperature magnetic moments for iron(II) Tutton salt and iron(II) fluorosilicate, both crystals containing $Fe(H_2O)_6^{2+}$ octahedral cations. Moments for $Fe(H_2O)_6.SiF_6$ and $Fe(H_2O)_6.(NH_4)_2(SO_4)_2$ are $5.5\mu_B$ and $5.2\mu_B$. The fluorosilicate crystallizes in a trigonal space group and any distortion in these complex cations is rigorously axial. The Tutton salt crystallizes in a monoclinic space group leaving the complex ions with no more than a centre of symmetry. Bond lengths favour a description of the Tutton salt geometry as predominantly tetragonal, although Mössbauer work reveals lower symmetry than axial for the electric field gradient. Originally, however, the distortion was taken to be tetragonal. The earliest successful treatment of the magnetic properties of these compounds considered the simultaneous effects of an axial crystal-field and spin–orbit coupling perturbations on the parent cubic field ground term $^5T_{2g}$. In this basis, the nature of the axial distortion is irrelevant, calculations in trigonal (D_{3d}) or tetragonal (D_{4h}) quantization giving identical results. The axial field is then adequately parameterized in terms of a $^5E-{}^5B_{2g}(D_{4h})$ or $^5E-{}^5A_{1g}(D_{3d})$ splitting Δ. The energy levels and wavefunctions in these systems were thus parameterized by Δ and λ, the spin–orbit coupling coefficient. A third parameter, k, the orbital reduction factor appearing in the effective magnetic moment operator

$$\mu = \mu_B(kL+2S) \tag{4.17}$$

allowed for the admixture of ligand p orbitals into the metal t_{2g} set, as is discussed elsewhere.[41] This three-parameter model, involving Δ, λ, k, had enjoyed considerable success in fitting observed magnetic moments throughout the transition block.[c] The expected result of using this approach on the iron(II) complexes was, of course, that both $Fe(H_2O)_6^{2+}$ ions should share common 'chemical properties', i.e. λ and k values, the different magnetic moments being reproduced by different Δ values. Surprisingly, however, widely different k values were deduced, being *ca.* 1.0 for the fluorosilicate and *ca.* 0.7 for the Tutton salt. In terms of the interpretation of k values then current, but since clarified, this meant no π-delocalization in the former $Fe(H_2O)_6^{2+}$ ion, and some 30 % in the latter!

Building on some earlier experience with magnetic properties[32, 28] of tetrahedral copper(II) ions in which the importance of including higher cubic-field terms in the calculations had been demonstrated, it was hoped to rationalize the iron(II) situation by expanding the basis set of functions to include the complete spin-quintet manifold. The impor-

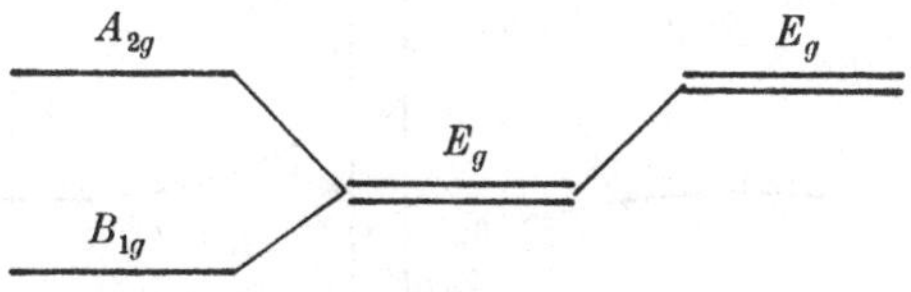

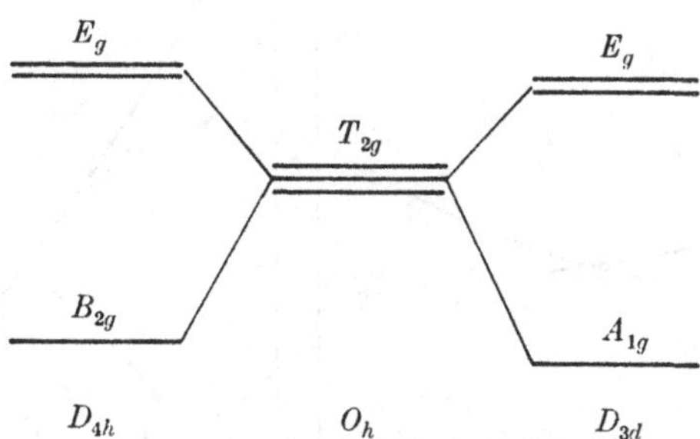

Figure 4.3. Splitting diagram for tetragonal and trigonal distortions of octahedral iron(II) systems.

tance of the (spin-doublet) E_g term in the copper calculations derived from mixing due to the spin–orbit coupling perturbation. If the same were true for the (spin-quintet) E_g term for the iron(II) compounds, calculated moments may be altered but we might expect no differentiation between the fluorosilicate and Tutton salts. The distinction between these ions arises by virtue of their different symmetries. The level splittings and designations in D_{4h} and D_{3d} symmetries are shown in figure 4.3, from which we observe that symmetry allows $E_g - E_g$ mixing *via* the axial crystal-field in the trigonal geometry while there can be no such low-symmetry field mixing between components of the cubic-field E_g and T_{2g} terms in the tetragonal case. The suggestion was, therefore, that axial crystal-field mixing, additional to spin–orbit mixing, might be responsible for the different magnetic properties of the two iron salts. The only problem was, however, that the magnitude of the distortions actually present in the fluorosilicate, as defined by X-ray-determined O—Fe—O bond angles, was less than 1° i.e. octahedral within experimental error. Could so small a distortion give rise to significant magnetic differences?

Using a point-charge crystal-field model as represented above, particularly by (4.14) and (4.15), energy levels were calculated as

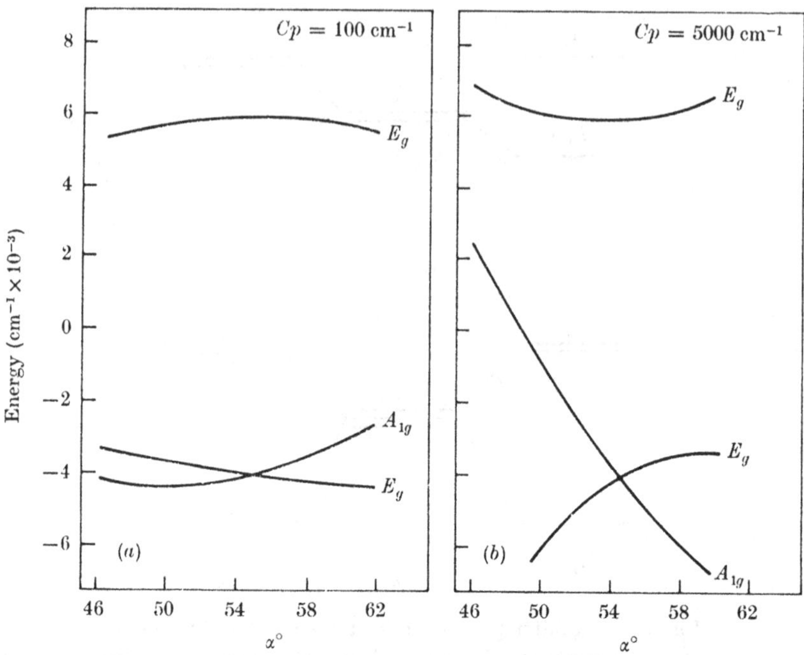

Figure 4.4. Energies of spin-quintet terms as functions of distortion angle for (*a*) small and (*b*) large *Cp* values: $Dq = 1000\,\text{cm}^{-1}$.

functions of *Cp* and α (and *Dq*). Figure 4.4 shows eigenvalues of all spin-quintet levels in D_{3d} symmetry as functions of α for two representative values of *Cp*. Energies are clearly dependent on *Cp* at least for such large variations as shown in figure 4.4. In particular, the ordering of the ground $^5T_{2g}$ components is reversed as *Cp* changes from 100 to $5000\,\text{cm}^{-1}$. Figure 4.5 demonstrates this in detail and shows how an accidental degeneracy of A_{2g} and E_g terms can occur even for a non-zero distortion (i.e. $\alpha \neq \alpha_{\text{oct}}$). In figure 4.5 (*b*) we see how, for $Cp = 1000\,\text{cm}^{-1}$, an E_g ground term persists throughout the angular range, irrespective of the sense of distortion, i.e. compression or elongation of the octahedron. This reversal in behaviour of the ground term components derives from the opposed effects of the second- and fourth-order harmonics in the crystal-field potential and when *Cp* takes large values, the second-order terms predominate. Even at this stage in the discussion we can see how an appreciation of the order-of-magnitude of *Cp* is important. Thus, for thermal population of one ground term component only, a single spin-allowed spectral band

$$(^5E_g \rightarrow {}^5E_g \quad \text{or} \quad {}^5A_{1g} \rightarrow {}^5E_g)$$

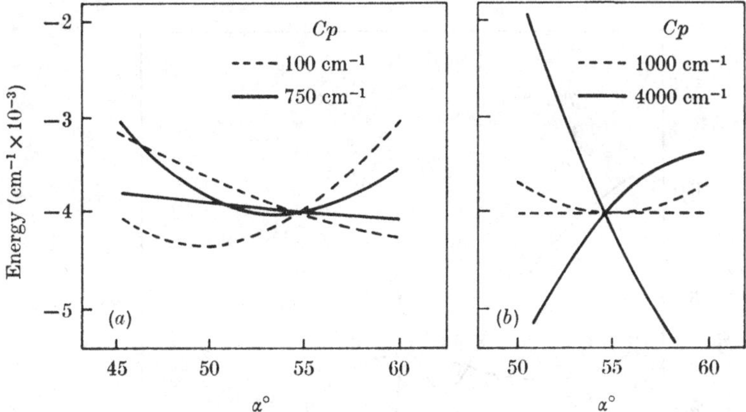

Figure 4.5. Energies of components of $^5T_{2g}$ term as functions of α for (*a*) small and (*b*) large *Cp* values: $Dq = 1000\,\text{cm}^{-1}$.

is expected. If *Cp* is small, corresponding to figure 4.4(*a*), this band frequency corresponds quite closely to $10Dq$, as for an exactly octahedral molecule. But for large *Cp* values, as in figure 4.4(*b*), an assignment of a distorted molecule as if it were regular could substantially overestimate *Dq*.

For magnetic properties, one requires eigenvalues derived from the perturbation including spin–orbit coupling. As a discussion of these does not clarify our present thesis we pass on to the magnetic moments themselves. In figure 4.6 are shown room-temperature mean moments calculated for trigonally-distorted octahedral d^6 complexes, plotted once more as functions of *Cp* and α for a fixed (typical) value of *Dq*. The angular dependences of mean moments are fairly gentle functions of α when *Cp* is small but can be very sensitive ones when *Cp* is large. Notice particularly how, for large *Cp* values, moments fall rapidly as the octahedron is compressed *slightly*. For example, with $Cp = 5000$ cm^{-1}, $\bar{\mu}$ falls from *ca.* 5.72 to *ca.* $5.35\mu_B$ as α increases by only 1° from the octahedral value (54.74°). It is just this behaviour which is considered responsible for the primary difference in mean moments of the fluorosilicate and Tutton salts of iron(II). Thus magnetic moments in this system can be significantly affected by trivially small angular distortions, provided *Cp* is a large quantity. The detailed analysis of the iron(II) problem (*q.v.*) did not, unfortunately, yield unambiguous results. This is frequently the case in such studies and reflects limitation of data as much as doubt about the validity of the model. It is likely that *Cp* is at least greater than 2000 cm^{-1}: fits are ever better for

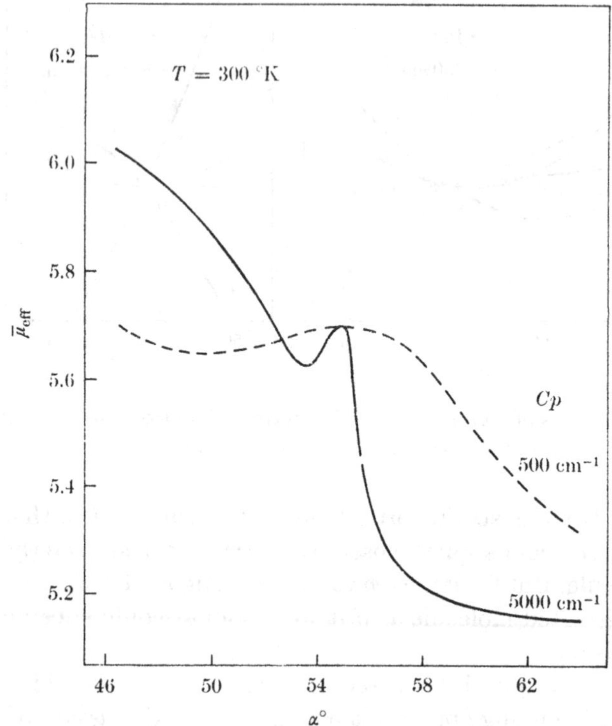

Figure 4.6. Average magnetic moments at 300 °K as functions of α for small and large Cp values: $Dq = 1000\,\text{cm}^{-1}$.

$Cp > 5000\,\text{cm}^{-1}$, corresponding to a ratio $Cp/Dq > 5$. With current information there seems little way of improving this estimate except by intelligent comparison with the results of similar experiments on other systems.

The situation for trigonally-distorted octahedral cobalt(II) complexes is even more interesting.[42] The schematic splitting diagrams for the spin-quartet levels in d^7 complexes under octahedral and D_3 rotation groups are shown in figure 4.7. Detailed calculations of the crystal-field splittings in D_{3d} symmetry using the same model and parameters as discussed for trigonal iron(II) systems have been performed and some typical results are shown in figures 4.8 and 4.9. The behaviour of energy levels varies markedly with the magnitude of Cp and we again observe reversal in the splitting of the ground ${}^4T_{1g}$ term as Cp increases. Analogous to the iron(II) case we see from figure 4.9, how an A_{2g} term lies lowest regardless of the sense of trigonal distor-

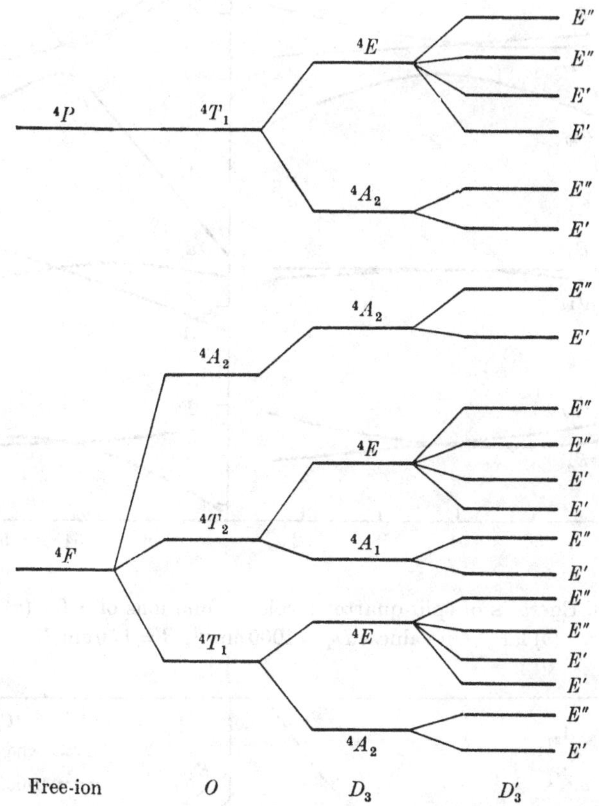

Figure 4.7. Splitting diagram for d^7 ions in O and D_3 rotation symmetry, and in D_3' rotation double group.

tion, when Cp is 3000 cm^{-1}. The angular sensitivity of levels becomes extremely marked for large Cp values and might be expected to have an important bearing on the interpretation of absorption spectra in such molecules.

The theoretical behaviour of average moments (shown at 300 °K in figure 4.10) for the cobalt(II) complexes as functions of α and Cp are similar to those for iron(II) in figure 4.6. Thus we observe moderate insensitivity of moments with respect to distortion angle for low Cp values (< 3000 cm^{-1}, say) but that this changes for higher Cp values when geometrical distortion can become the dominant feature. The absolute magnitudes of these mean moments are dependent on several factors, spin–orbit coupling, Dq, k, etc., of course, but the magnitude of Cp clearly takes on an important role in establishing this relative

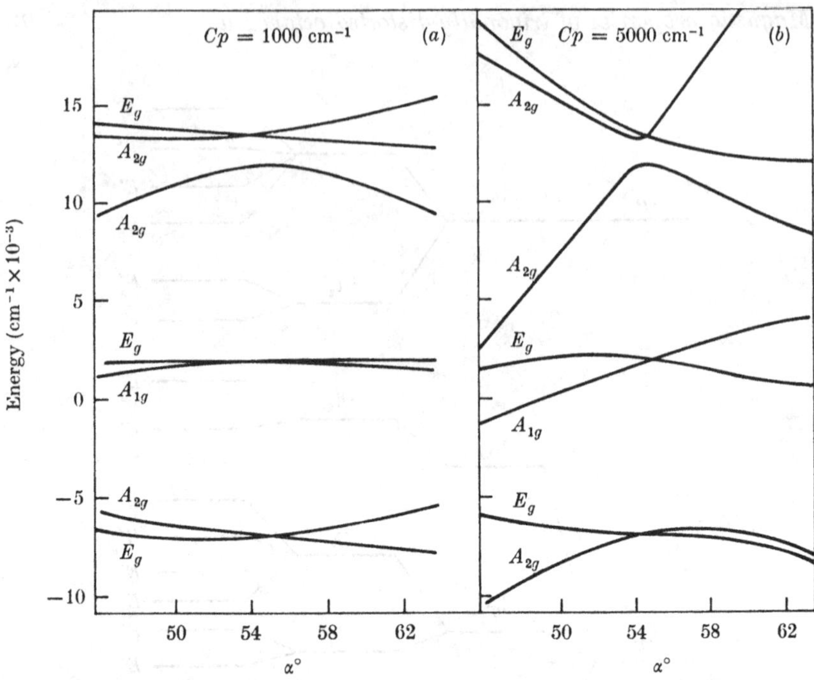

Figure 4.8. Energies of spin-quartet levels as functions of α for (a) small and (b) large Cp values: $Dq = 1000\,\text{cm}^{-1}$, $B = 850\,\text{cm}^{-1}$.

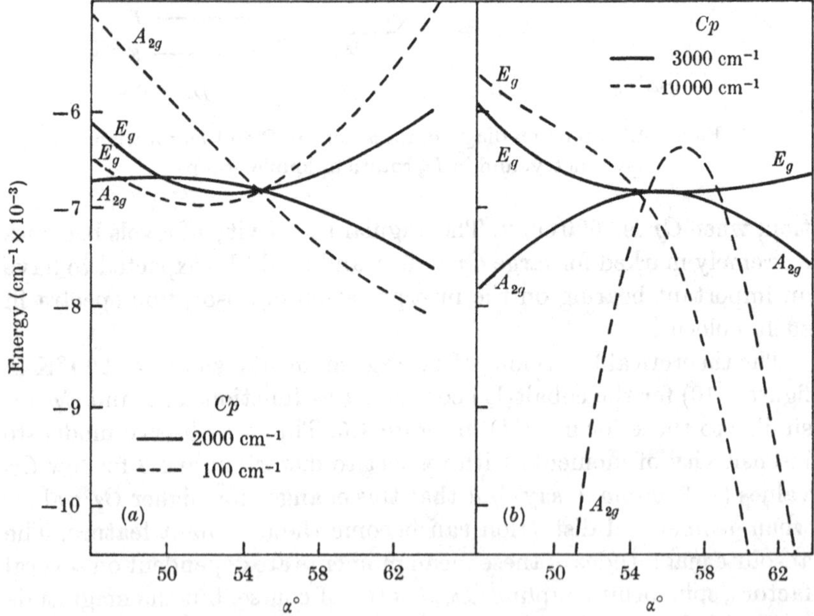

Figure 4.9. Energies of components of $^4T_{1g}$ term as functions of α for (a) small and (b) large Cp values: $Dq = 1000\,\text{cm}^{-1}$, $B = 850\,\text{cm}^{-1}$.

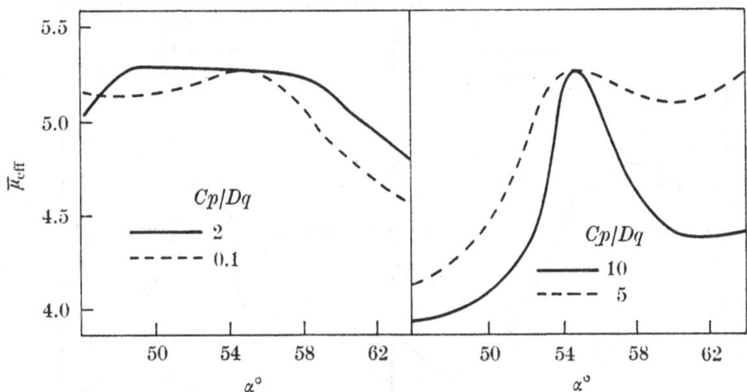

Figure 4.10. Average magnetic moments as functions of α: $T = 300\,°K$, $Dq = 1000\,cm^{-1}$, $B = 850\,cm^{-1}$, $\lambda = \lambda_0 = -172\,cm^{-1}$, $k = 1.0$.

sensitivity to geometrical distortion or to wavefunction character (as defined, say, by the orbital reduction factor, k). Perhaps the most striking aspects of these calculations concerns the magnetic anisotropy. Figure 4.11 shows plots of $\Delta\mu$ (being the difference in molecular magnetic moment parallel and perpendicular to the axis of distortion) as functions of α for a family of Cp values.

When Cp is a small quantity (figure 4.11 (a)), calculated anisotropies follow a fairly 'normal' pattern in that the sign of anisotropy (i.e. the relative size of $\mu_{\parallel}$ and $\mu_{\perp}$) depends upon the sense of distortion. In this particular case a trigonally-squashed octahedron has a larger magnetic moment perpendicular to the molecular triad than parallel to it. The magnitude of $\Delta\mu$ tends to maximize for fairly small distortions of either sign from octahedral and then either decreases to zero, when there is an orbitally non-degenerate ground term ($\alpha > \alpha_{oct}$ in figures 4.11 and 4.9), or attain a constant value for a doubly degenerate ground term ($\alpha < \alpha_{oct}$ in figures 4.11 and 4.9). When Cp takes large values, however, the behaviour of anisotropies with distortion angle becomes most complex. To quote the particular case for $Cp = 10\,000\,cm^{-1}$ in figure 4.11 (b) we note that as distortion varies smoothly from a trigonal elongation to a squash, $\Delta\mu$ changes from essentially zero ($\alpha < 52°$); to $-1.0\,\mu_B$ at α *ca.* 54.0°; to identically zero at the isotropic, octahedral situation when $\alpha = 54.74°$; to $+1.0\,\mu_B$ at α *ca.* 56°; zero at 58°; negative ($58° < \alpha < 63°$); then positive again! This 'switch-back' behaviour of anisotropy clearly rules out simple correlation with the geometrical sense of distortion and, incidentally, shows again how such

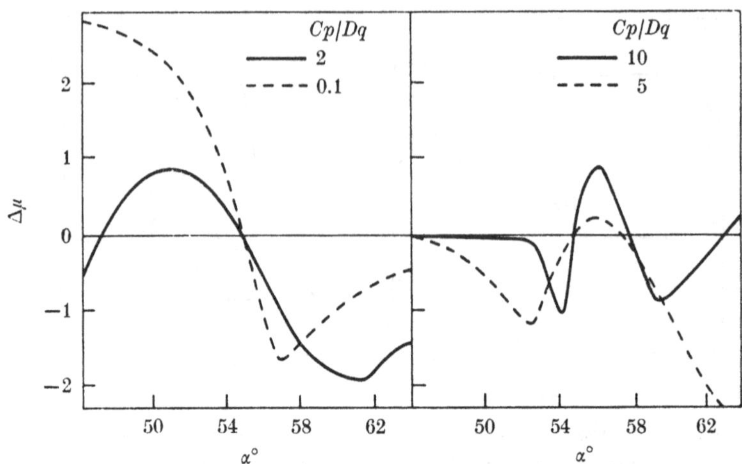

Figure 4.11. Magnetic anisotropies ($\Delta\mu = \mu_{\parallel} - \mu_{\perp}$) as functions of α: $T = 300$ °K, $Dq = 1000\,\text{cm}^{-1}$, $B = 850\,\text{cm}^{-1}$, $\lambda = \lambda_0 = -172\,\text{cm}^{-1}$, $k = 1.0$.

a complex with zero magnetic anisotropy need not be structurally isotropic i.e. regular octahedral. An understanding of the detailed magnetic properties of systems such as these obviously requires, or perhaps will give, information on the magnitude of the second-order crystal-field radial parameter *Cp*.

Measurements of the powder susceptibilities and single-crystal magnetic anisotropies over a temperature range for two trigonally-distorted octahedral cobalt(II) compounds have been compared with this model:[42] cobalt fluorosilicate, $Co(H_2O)_6^{2+}SiF_6^{2-}$, and hexakis (imidazole) cobalt(II) dinitrate, $Co(im)_6^{2+}(NO_3^-)_2$. The results were ambiguous to the extent that *Cp*/*Dq* ratios for the compounds were either less than about 2 or greater than 8. It does seem clear, however, that a *Cp*/*Dq* value between these limits is incompatible with the experimental results. It was found that if *Cp* takes small values then $Cp/Dq \lesssim 2.0$ for both compounds, or if not we can construct the following table:

	Dq (cm^{-1})	Cp (cm^{-1})	Cp/Dq	
$Co(im)_6(NO_3)_2$	1130	8000–12000	7.1–10.6	(4.18)
$Co(H_2O)_6.SiF_6$	900	12000–16000	13.3–17.8	

Thus for the higher range for *Cp*, we find the larger *Cp*/*Dq* ratio is associated with the smaller *Dq* value.

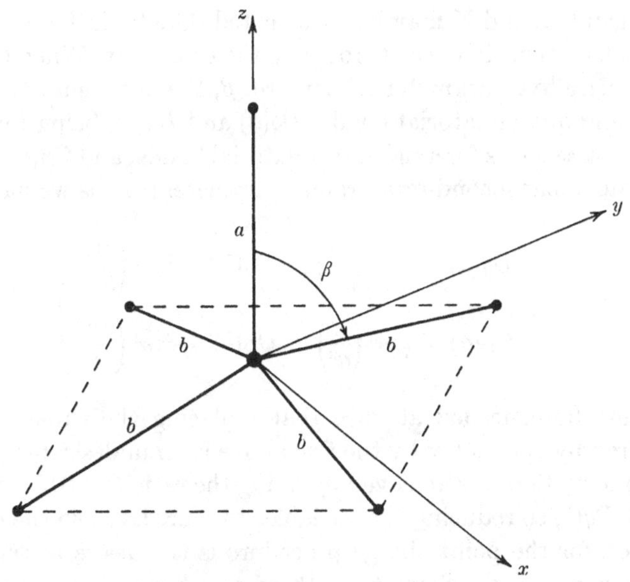

Figure 4.12. Axes and positions of ligands in five-coordinate, C_{4v} symmetry.

EXAMPLES FROM SPECTRA

We have discussed some examples in which *Cp* is an important parameter in understanding magnetic properties: we now consider its relevance to electronic spectra. The first example we choose here concerns the quite different molecular geometry of a square-pyramidal five-coordinate cobalt(II) complex.[33,34] The compounds $(Ph_2MeAsO)_4$ $MX^+.Y^-$, where M = Co, Ni, Mn and X and Y are ClO_4 or NO_3 and may be the same or different, crystallize isomorphously in a tetragonal space group with molecular symmetry C_4. (This has to be achieved strictly, by disordering or free-rotations of the X and Y groups.) The molecular geometry, involving four basal arsine oxide groups and an axial perchlorate or nitrate, might be approximated as having C_{4v} local symmetry for the chromophore, as in figure 4.12. Symmetry considerations alone show that a crystal-field potential in this symmetry has three terms:

$$V_{C_{4v}} = aY_2^0 + bY_4^0 + c(Y_4^4 + Y_4^{-4}). \qquad (4.19)$$

Matrix elements are commonly parameterized in terms of the 'quadrate' coefficients *Dq*, *Dt* and *Ds* as discussed in the next chapter, but in an attempt to separate angular from radial factors, in a fashion similar to that already discussed, a point-charge formalism may be used. A table of r', θ', ϕ' coordinates for point-charges representing the ligands L and X may be constructed (like table 4.1, for example) and calculations followed through in the usual way. When this is done there arise five parameters. They are: β, the angle subtended by the axial and any equatorial bonds; $Dq(a)$ and $Dq(b)$, being fourth-order radial parameters for axial and equatorial bonds; and $Cp(a)$ and $Cp(b)$, the equivalent second-order radial parameters. Thus we have:

$$\left.\begin{aligned} Dq(a) &= \tfrac{1}{6}ze^2\left(\frac{\overline{r^4}}{a^5}\right), \quad Dq(b) = \tfrac{1}{6}ze^2\left(\frac{\overline{r^4}}{b^5}\right), \\ Cp(a) &= \tfrac{2}{7}ze^2\left(\frac{\overline{r^2}}{a^3}\right), \quad Cp(b) = \tfrac{2}{7}ze^2\left(\frac{\overline{r^2}}{b^3}\right). \end{aligned}\right\} \tag{4.20}$$

It must be conceded at once that replacing the three parameters required by symmetry by the five above is an undesirable procedure. Some reduction was achieved by fixing the ratio $Cp(a)/Dq(a)$ equal to $Cp(b)/Dq(b)$ so reducing the variables to four. Even so the only justification for the point charge procedure is the desire to separate the angular parameter from the others and hence hopefully compare Cp and Dq values with those obtained for other molecules and geometries.

An important point about parameterization arises here. It might be supposed that the angle β in figure 4.12 (or, indeed, α in the trigonal models) need not be regarded as a variable but as fixed by the known molecular geometry. While this procedure has been followed by some authors we do not do so for the following reasons. The point-charge model seeks to represent the donor atom, or worse still a polyatomic ligand, as a point-charge (or a point-dipole) so that the bond direction as defined by the internuclear vectors M–L may not faithfully reproduce the *effective* direction of maximum influence. Any lone pairs on the donor atom may also tend to invalidate such an assumption. Even so it might be supposed that, given all the other approximations this model involves, the assumption of equivalence between effective ligand-field directions and internuclear vectors might be a fair one. However, as demonstrated for the trigonally-distorted octahedral molecules above and as also transpires for the present square pyramidal complexes, the calculated sensitivity of energy levels and magnetic moments to the angular parameters can be very great: a fraction of one degree can be significant. In view of this, we always regard angular factors as *parameters* of the system to be fitted to experiment. A reasonable agreement with X-ray-determined bond angles is to be expected, however.

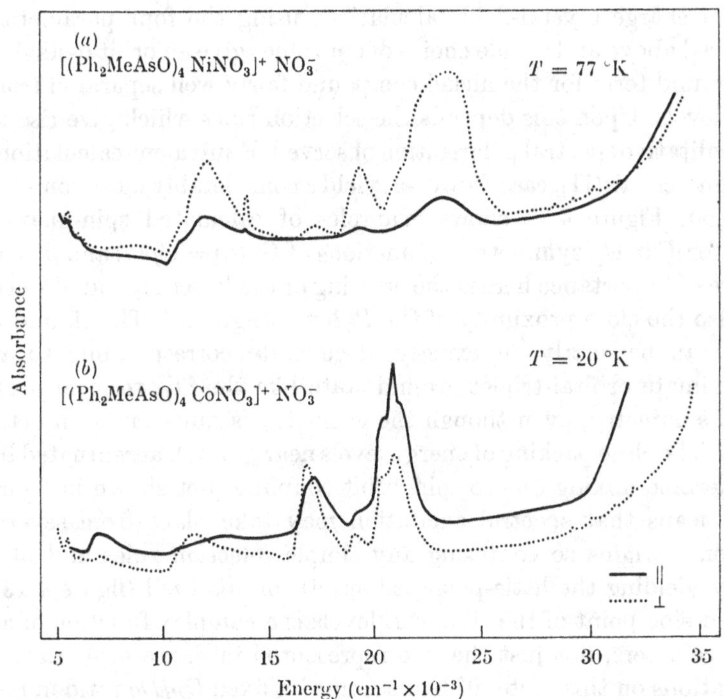

Figure 4.13. Polarized crystal spectra of $[(Ph_2MeAsO)_4MNO_3]^+NO_3^-$.

Returning to the five-coordinate complexes we wish to illustrate two points – one about the magnitudes of the radial parameters and one about their effect on spectral polarizations. Polarized spectra and single-crystal moments were measured for $(Ph_2MeAsO)_4MX_2$ where M = Co and Ni and X = NO_3^-. The spectra[33, 34] are shown in figure 4.13. Crystals of the nickel compound are strongly dichroic, appearing red with light polarized perpendicular to the four-fold molecular axis and yellow for light polarized parallel. This is shown clearly by the essentially perpendicularly polarized spectrum given in figure 4.13(*a*). By contrast, the spectrum of the cobalt analogue is hardly polarized at all. One might naïvely suppose that in the nickel compound there is a strongly asymmetric ligand field but that the cobalt one is nearly symmetric (pseudo-octahedral?). However, the magnetic properties reveal a quite different situation in that large anisotropies are found for both complexes:

$$L_4CoX_2: \mu_\parallel = 3.95, \quad \mu_\perp = 5.54\mu_B \quad \text{at} \quad 300\,°\text{K};$$

$$L_4NiX_2: \mu_\parallel = 3.11, \quad \mu_\perp = 3.52\mu_B \quad \text{at} \quad 300\,°\text{K}.$$

Point-charge crystal-field calculations using the four parameters discussed above and a wide choice of Cp values give an orbital-singlet, $^3B_1$, ground term for the nickel compound fairly well separated from other levels. Upon this depends the selection rules which give rise to the qualitative spectral polarization observed. Equivalent calculations for the d^7, cobalt(II), case, however, yield a considerably more complex situation. Figure 4.14 shows examples of calculated spin-quartet levels for d^7 in C_{4v} symmetry as functions of $Cp(a) = Cp(b)$ and β. The feature of importance here is the crossing of the lower A_2 and E levels and also the close proximity of the B_2 level to ground. The A_2 and E levels can be nearly, or exactly, degenerate corresponding to an approximate orbital-triplet ground state like the $^4T_{1g}$ term in octahedral symmetry, even though the geometry is quite far from octahedral. The close packing of energy levels near ground, accentuated by considerable mixing due to spin–orbit coupling (not shown in figure 4.14) means that spectral excitation may take place from several electronic origins so confusing any simple selection rules and ultimately yielding the little-polarized spectrum observed (figure 4.13). The crossing point of the A_2 and E levels is a complex function of all the parameters, not just the two represented in figure 4.14. Earlier calculations on this system[13] had assumed a fixed Cp/Dq ratio in line with some theoretical calculations of Ballhausen and Ancmon,[4] which we discuss in chapter 7. The choice not to regard Cp as a parameter of the system failed to reveal the ambiguity in the nature of the ground term, particularly as it was also assumed[13] that β should be fixed at the value of the X-ray determined bond angle. The large magnetic anisotropy observed derives from the non-cubic nature of the level splitting, involving slight splitting of the A_2 and E levels (anisotropy tends to be largest for very small splittings) and also the close proximity of the B_2 term to ground.

The relevance of this discussion of these five-coordinate complexes to our general thesis is that it is possible to compare parameters associated with molecules of widely differing geometry and that the second-order radial parameters are again an important factor. Results from the detailed analyses of these systems suggested the ratio Cp/Dq lies in the range 2–4.

Norgett, Thornley and Venanzi[75] studied the spectra of some five-coordinate complexes with d^6, d^7 and d^8 configurations. They were obliged to simplify the situation by assuming a higher symmetry than the molecules actually possess and to take similar radial parameter values for axial and equatorial ligands in these essentially C_{3v} systems.

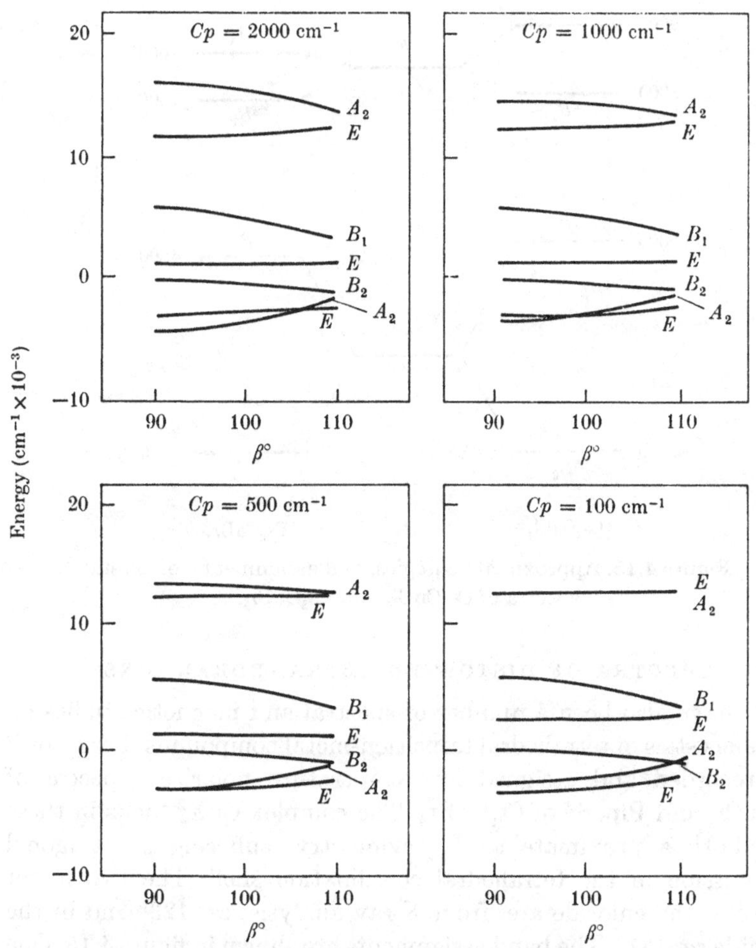

Figure 4.14. Energies of spin-quartet levels as functions of β and Cp: $Dq = 600\,\text{cm}^{-1}$, $B = 850\,\text{cm}^{-1}$, Cp/Dq (axial) $= Cp/Dq$ (basal).

Noting that their parameters Q_2 and Q_4 are related to Cp and Dq by:

$$Q_2 = \tfrac{7}{2}Cp, \quad Q_4 = 6Dq,$$

their results give:

	Q_2/Q_4	Cp/Dq
d^6, Fe^{2+}	1.3	2.2
d^7, Co^{2+}	2.3	3.9
d^8, Ni^{2+}	2.0	3.4

and we note that Cp/Dq ratios again lie in the range 2–4.

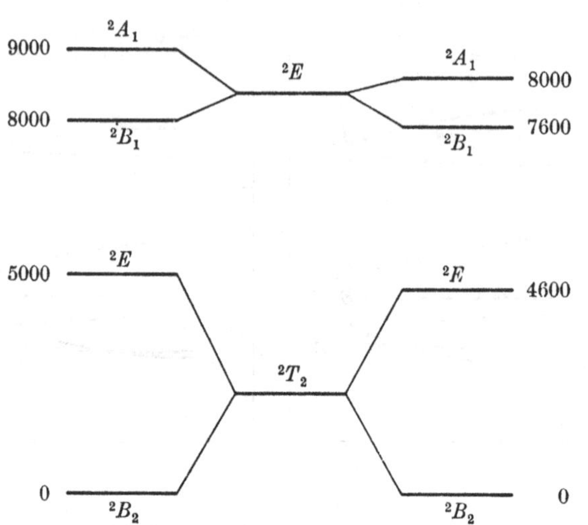

Figure 4.15. Approximate energies, and assignments, of crystal spectra of Cs_2CuCl_4 and Cs_2CuCl_5.

SPECTRA OF DISTORTED TETRAHEDRAL IONS

There have also been a number of spectral and magnetic studies on single-crystals of tetrahedral transition-metal compounds. Ferguson[24] has recorded and assigned low-temperature polarized spectra of Cs_2CuCl_4 and Piper[84] of Cs_2CuBr_4. The complex CuX_4^{2-} ions in these salts both approximate to D_{2d} symmetry, suffering a tetragonal compression of the tetrahedral coordination shell. The two larger angles in the chloride are, from X-ray analysis, *ca.* 123° and in the bromide *ca.* 137°. The band assignments are shown in figure 4.15. One may fit these band positions to the D_{2d} point-charge model described at the beginning of this chapter by variation of α, Dq and Cp. It is found that best fits for each complex occur for 2α values which are very close to the crystallographic values quoted above. Neglecting spin–orbit coupling the following values are obtained for Dq and Cp/Dq:

	Cs_2CuCl_4	Cs_2CuBr_4	
Dq	1150	900 cm^{-1}	(4.21)
Cp/Dq	1.76	1.54	

The first point to note is that the Dq values quoted correspond to the definition of Dq in (2.69) so that those expecting much smaller values should regard these values as $\frac{9}{4}\Delta_{tet}$, where Δ_{tet} is the more usual measure.

The Cp/Dq ratios for these tetrahedral copper(II) compounds appear to be much lower than those discussed for other geometries above. The inclusion of spin–orbit coupling into the fitting procedure makes some difference but does not substantially alter the low ratio given in (4.21).

Couch and Smith[18] have reported studies of the electronic absorption spectrum of tetragonally-distorted $NiCl_4^{2-}$ doped in a Cs_3MgCl_5 host lattice. Extremely fine *ca.* 5 °K spectra have been recorded and the energies of seven transitions were hopefully established, partly by investigation of their vibrational fine structure and partly by recourse to crystal-field calculations which included spin–orbit coupling effects. These transition energies were fitted by least-squares adjustment of the six parameters – Racah's interelectron repulsion factors B and C, the spin–orbit coupling coefficient, and three crystal-field parameters B_0^2, B_0^4 and B_4^4. These latter may be related to the point-charge formalism used above. When this is done, we find (*a*) $Dq = 734\ \text{cm}^{-1}$ (i.e. conventional Δ_{tet} for the 'tetrahedron' $= \frac{4}{9} \times 7340 = 3270\ \text{cm}^{-1}$), (*b*) α, the angle between any bond and the S_4 axis is 53.2°, as compared with 53° in Cs_3CoCl_5 (from X-ray analysis[29]), and (*c*) $Cp = 4790\ \text{cm}^{-1}$. Thus Couch and Smith find $Cp/Dq \sim 6.5$, much larger than the corresponding values for the tetrahedral copper(II) complexes. The values of most parameters deduced from single-crystal spectral or magnetic properties are usually subject to error due either to inaccuracies in data or assignment or simply to an insensitivity of the theoretical model to particular parameters. We have experienced all these problems in our own research, of course. In the present case of $NiCl_4^{2-}$ ion spectra there may be some reason to doubt the Cp/Dq ratio deduced above. The relevant parameter B_0^2 is fixed only by two splittings. One splitting is associated with components of the ^1D(free-ion) term and one with components of the $^3T_1(P)$ (T_d) term. In both cases, the former because of an intensity distribution, the latter because of the appearance of an 'extra' peak in the band, the Jahn–Teller effect has been invoked. The band system which most determines B_0^2 is the $^3T_1(P)$. The authors wished to assign transitions to two E states (D_{2d}) to two of three peaks in the perpendicular polarized spectrum – figure 4.16. The lowest energy band should correspond to the E state with the most *orbital* degeneracy (as opposed to spin) and be most subject to Jahn–Teller distortion.[18] They ascribe the middle of the three bands (2) to a Jahn–Teller component of this and regard the

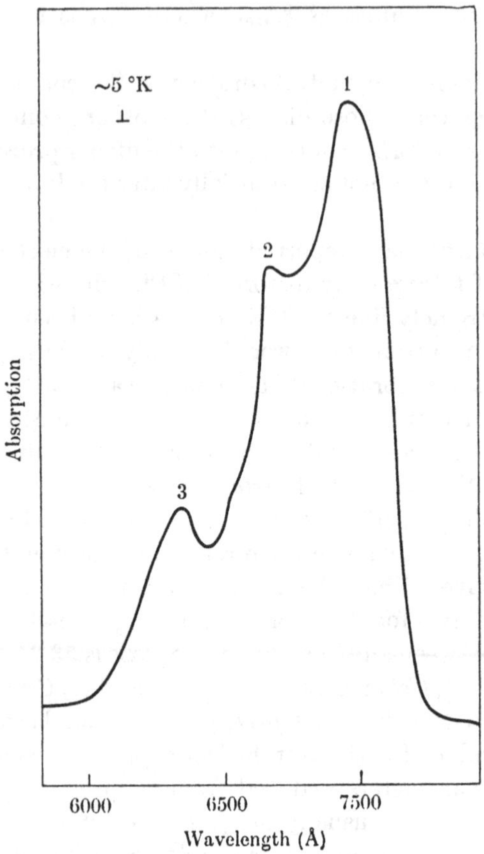

Figure 4.16. Detail of perpendicular polarized spectrum $[^3T_1(P)]$ of $NiCl_4^{2-}$ in Cs_3MgCl_5.

energy of the lower energy component (1) of this pair of bands to the origin of the transition. This in itself may be questioned, but of even more concern here is the possibility of assigning the highest energy peak (3) to this Jahn–Teller pair, rather than the middle (2). If that were done the $E^1 \leftrightarrow E^2$ splitting would be 630 cm^{-1} or less, if a 'centre of gravity' correction for the Jahn–Teller components (1) and (3) were made, rather than 1500 cm^{-1}. From figure 6 in Couch and Smith's paper, a rough estimate of B_0^2 would then be -900 cm^{-1} (say) rather than -2280 cm^{-1}, giving a resulting Cp/Dq ratio of 2.7. The 'accuracy' of the fitting of theory to experiment in these authors' paper need not repudiate this suggestion: their theoretical prediction of the energies of the ^1G components are relatively poor.

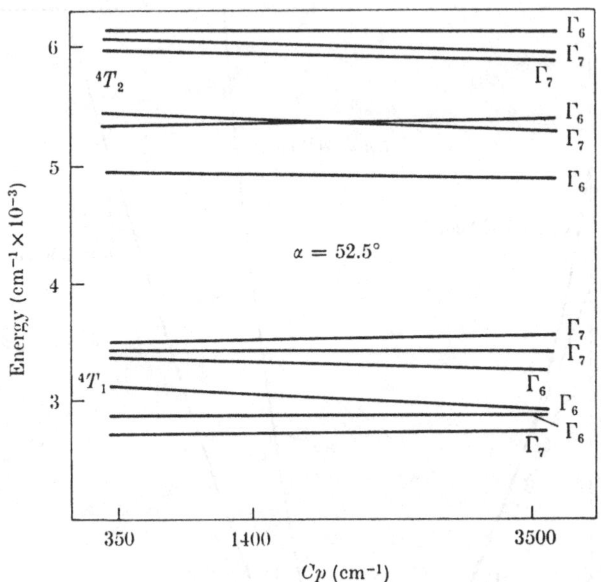

Figure 4.17. Calculated energies of $^4T_1$ and $^4T_2$ components as functions of Cp:Dq = 700 cm^{-1}, ζ = 515 cm^{-1}.

We do not wish to declare that the original assignments are incorrect or even that they are unlikely to be so. We merely want to show how, even for experimental work as fine as that on $Cs_3(Ni/Mg)Cl_5$, doubts about the accuracy of some parameter values can arise. Perhaps we might summarize by taking a Cp/Dq value of 6.5 for $NiCl_4^{2-}$ ions but acknowledge an uncertainty in this figure. Again, one reason for gathering together current ideas and information on ligand-field parameters in this book is to try to establish an empirical and perhaps semi-theoretical series of these quantities, like the Spectrochemical Series, so that a value of Cp/Dq like that for $NiCl_4^{2-}$ may be recognized as 'typical' or 'unusual' and, if appropriate, examined further.

SECOND-ORDER TERMS ARE NOT ALWAYS VERY IMPORTANT

Magnetic moments and spectra are not always sensitive to Cp. Tetragonally-distorted tetrahedral cobalt(II) ions furnish an interesting example of this.[37] The susceptibility and g-value anisotropies of Cs_3CoX_5 (X = Cl, Br) have been studied many times and these studies have recently been supplemented by some of the CoX_4^{2-} ions in another lattice, *viz.* $(Et_4N)_2CoX_4$. The insensitivity of energy levels

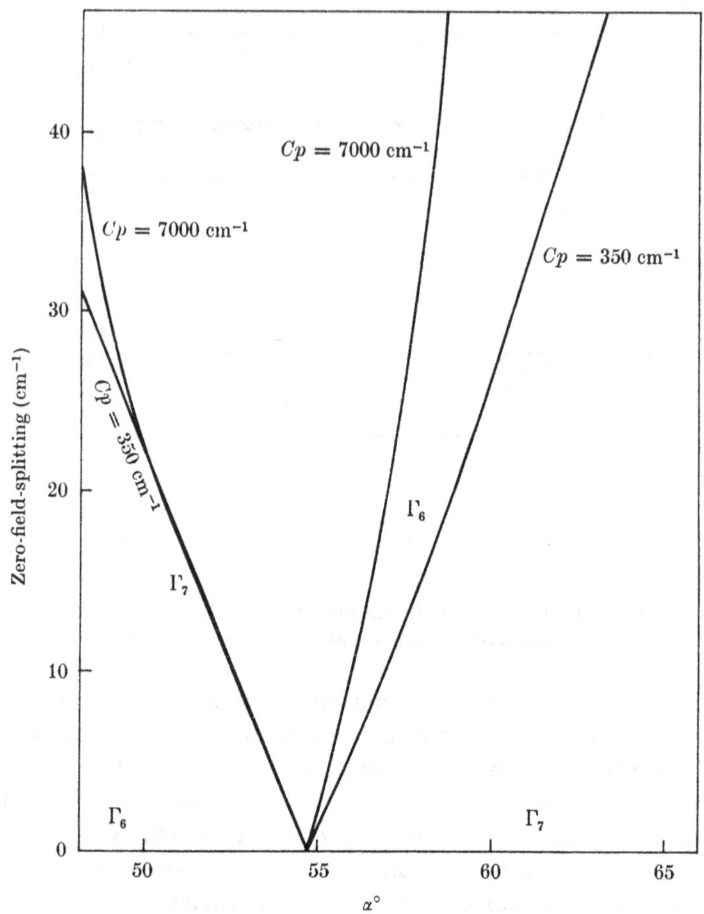

Figure 4.18. Zero-field-splitting as function of distortion angle and Cp: $Dq = 700\,\text{cm}^{-1}$, $\zeta = 515\,\text{cm}^{-1}$.

and magnetic properties to the second-order term in the crystal-field potential Y_2^0 was first demonstrated by Jesson.[55] In the point-charge formalism discussed above, the coefficient of Y_2^0 involves both α and Cp so that Jesson's work does not necessarily imply insensitivity of observables to Cp as there might be a cancellation effect between functions of α and Cp. Recent work, however, does show that for small distortions, appropriate to the observed geometries of tetrahedral cobalt(II) halides, all observables (at least connected with spin-quartet levels) are extremely insensitive to Cp. Figure 4.17 shows energy levels from spin-quartet levels of d^7 in D_{2d} symmetry as functions of Cp for $\alpha = 52.5°$. The mean magnetic moments of these systems are quite well reproduced by the equation

$$\mu = \mu_{\text{spin-only}}\left(1-\frac{4k^2\lambda}{\Delta_{\text{tet}}}\right). \tag{4.22}$$

As the '$^4A_2-{}^4T_2$' splitting is insensitive to Cp, so are calculated mean moments. Magnetic anisotropies depend, amongst other things, upon the magnitude of the zero-field-splitting (z.f.s.) of the formal $^4A_2$ ground term. Figure 4.18 shows calculated z.f.s. as functions of α and Cp. From this we observe that, for small distortions, especially when $\alpha < \alpha_{\text{tet}}$ (i.e. tetragonal elongation, as happens to be appropriate for the compounds under consideration), zero-field-splittings are remarkably insensitive to the magnitude of Cp and, in consequence, so are magnetic anisotropies.

Magnetic properties of tetragonally-distorted tetrahedral nickel(II) compounds also appear to be fairly insensitive to Cp, as discussed elsewhere.[44,45,46] The behaviour of these nickel and cobalt systems does not appear to be a property of the tetrahedron, however, as similar behaviour was not observed for the 'one-electron' copper(II) system.

The systems discussed here have been chosen to illustrate the relevance or otherwise of Cp and to give some idea of its size. We can discern already that it is a larger quantity than Dq, the ratio Cp/Dq lying perhaps in the range 2–14. There is also some indication that the ratio increases with coordination number. Such conclusions are necessarily tentative at this stage as there have been relatively few single-crystal spectroscopic and magnetic studies from which Cp values may be deduced. We hope that the rationalization of low-symmetry ligand-field parameters we attempt in this book may encourage further experimental work in this direction.

In chapter 6 we shall discuss theoretical aspects and rationalizations of these parameters but before that we look at the parameters commonly used to describe systems suffering distortion other than angular. These predominantly involve the tetragonally-distorted octahedral geometry and the parameters Ds and Dt, for which considerably more experimental data exist.

GENERAL REFERENCES

(*a*) C. J. Ballhausen, *Introduction to Ligand Field Theory*, McGraw-Hill, New York, 1962.

(*b*) B. N. Figgis, *Introduction to Ligand Fields*, Interscience, New York, 1966.

(*c*) M. Gerloch and J. Lewis, Chemistry and Paramagnetism, *Revue de Chemie Minerale*, 1969, **6**, 19.

5
RADIAL PARAMETERS FOR QUADRATE SYMMETRY

DEFINITIONS OF Dt AND Ds

A particularly common distortion found in octahedral molecules involves the lowering of O_h symmetry to D_{4h}. This may be the result of lengthening or shortening two *trans* M–L bonds with respect to the other four in an otherwise regular octahedral molecule. More usually D_{4h} symmetry is achieved by a chemical dissimilarity in two *trans*-ligands from the other four as in *trans* MX_4Y_2 complexes. Using the same sort of group-theoretical arguments as at the beginning of the previous chapter, it is simple to show that, neglecting the spherical term Y_0^0, the potential in D_{4h} symmetry may be written as:

$$V_{D_{4h}} = aY_2^0 + bY_4^0 + c(Y_4^4 + Y_4^{-4}), \tag{5.1}$$

and is thus formally similar to that for the tetragonally-distorted tetrahedron with D_{2d} symmetry. However, the particular form of D_{4h} geometry in the octahedron invites an alternative parameterization scheme[a] as follows.

The total potential may be regarded as the sum of an octahedral and a tetragonal term, *viz.*

$$V_{D_{4h}} = V_{\text{oct}} + V_{\text{tetrag}}, \tag{5.2}$$

so that, subtracting

$$V_{\text{oct}} = Y_4^0 + (\sqrt{\tfrac{5}{14}})(Y_4^4 + Y_4^{-4}) \tag{5.3}$$

from $V_{D_{4h}}$ in (5.1), we get

$$V_{\text{tetrag}} = AY_2^0 + BY_4^0. \tag{5.4}$$

This form of V_{tetrag} shows how we may regard the tetragonal distortion as having affected the two axial ligands rather than the four equatorial ones. In other words the crystal-field effect of the four equatorial ligands in a tetragonally-distorted molecule is regarded as representative of an octahedral field V_{O_h} upon which V_{tetrag} is a perturbation.

When evaluating matrix elements of $V_{D_{4h}}$ we shall encounter Dq in connection with V_{oct}, of course, plus two other parameters associated with the terms Y_2^0 and Y_4^0 in V_{tetrag} of (5.4). As in the case of the tetragonal or trigonal angular distortions discussed in the previous chapter, these two parameters involve second- and fourth-order radial parameters, respectively. They are not Cp and Dq as defined earlier, however, as the distortions here do not involve angular changes but rather bond length or effective bond length changes. Instead, the parameters Ds and Dt are used and defined in the following open way:

$$\left.\begin{aligned} Ds &= \int_0^\infty R(d).f_2(r).R(d).r^2 dr, \\ Dt &= \int_0^\infty R(d).f_4(r).R(d).r^2 dr. \end{aligned}\right\} \tag{5.5}$$

The definitions are somewhat analogous to those in (2.66). Ds, which involves the product of the radial part of the metal d wavefunction squared and a function involving r^2 is associated with the Y_2^0 term in the potential V_{tetrag}: Dt involves r^4 and is associated with Y_4^0.

The single-electron matrix elements of a d-orbital basis in D_{4h} symmetry expressed in terms of these parameters are as follows:[a]

$$\left.\begin{aligned} \langle 0|V_{\text{tetrag}}|0\rangle &= -2Ds-6Dt, \\ \langle \pm 1|V_{\text{tetrag}}|\pm 1\rangle &= -Ds+4Dt, \\ \langle \pm 2|V_{\text{tetrag}}|\pm 2\rangle &= 2Ds-Dt. \end{aligned}\right\} \tag{5.6}$$

There are no non-zero off-diagonal terms as V_{tetrag} contains only harmonics Y_l^m with $m = 0$. To these must be added the matrix element of V_{oct} from (2.54), giving:

$$\left.\begin{aligned} \langle 0|V_{D_{4h}}|0\rangle &= 6Dq-2Ds-6Dt, \\ \langle \pm 1|V_{D_{4h}}|\pm 1\rangle &= -4Dq-Ds+4Dt, \\ \langle \pm 2|V_{D_{4h}}|\pm 2\rangle &= Dq+2Ds-Dt, \\ \langle \pm 2|V_{D_{4h}}|\mp 2\rangle &= 5Dq. \end{aligned}\right\} \tag{5.7}$$

Spectra and magnetic moments of ions with D_{4h} symmetry may now be interpreted within the framework of the three parameters Dq, Dt and Ds. It is important to note that these are symmetry parameters in that nothing barring the D_{4h} symmetry group was assumed in defining them (and, of course, the notion that we are dealing with a crystal-field potential): the form of the f functions in (5.5) is not required. As such they may be regarded as the 'most basic' crystal-field parameters in D_{4h} symmetry being the closest crystal-field approach to empirical or phenomenological parameters. The electronic properties of many compounds have been interpreted or expressed with their aid and some tables of experimental Ds and Dt values compiled. Some authors have

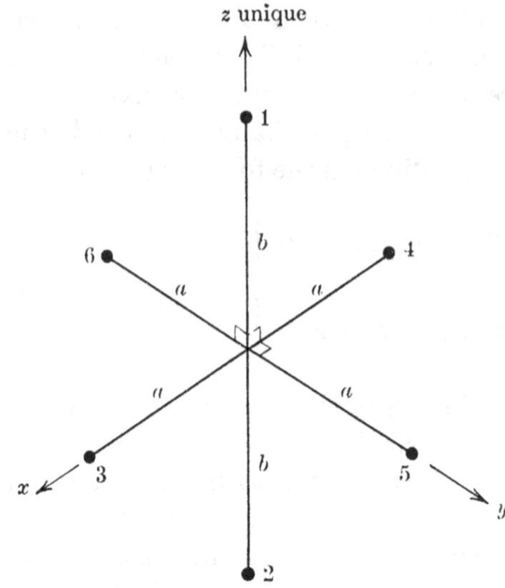

Figure 5.1. Axes and ligand positions in D_{4h} symmetry.

TABLE 5.1

Point	r'	θ'	ϕ'
1	b	0	0
2	b	π	0
3	a	$\frac{1}{2}\pi$	0
4	a	$\frac{1}{2}\pi$	π
5	a	$\frac{1}{2}\pi$	$\frac{1}{2}\pi$
6	a	$\frac{1}{2}\pi$	$\frac{3}{2}\pi$

also been interested in recording the observed ratios Ds/Dt which Perumareddi[82] defines as κ:

$$\kappa = Ds/Dt. \tag{5.8}$$

Although κ usually takes positive values, negative ones have been observed. A positive or negative sign for κ does *not* correspond to a tetragonal elongation or compression: remember that V_{tetrag} is the sum of second-order and fourth-order terms, (5.4). Perumareddi[82] has described Ds and Dt as being proportional to integrals $\overline{(r^2/a^3)}$ and $\overline{(r^4/a^5)}$, like Cp and Dq. If this were so, negative κ values, implying one or other of $\overline{(r^2/a^3)}$ or $\overline{(r^4/a^5)}$ is negative, would be difficult to understand, as these functions are intrinsically positive in a point-charge or dipole model as shown in the next chapter. In this way a negative κ value might be taken to imply the breakdown of the crystal-field model in some way with unspecified properties of molecular-orbital theory being required to save the day. None of this is necessary, however, as the identification of Ds and Dt as proportional to $\overline{(r^2/a^3)}$ and $\overline{(r^4/a^5)}$ is incorrect. This is best shown by recourse to the point-charge formalism.

RELATIONSHIPS BETWEEN *Dt*, *Ds* AND *Cp*, *Dq*

Figure 5.1 and table 5.1 describe the tetragonally-distorted octahedron within a point-charge model. The distortion is represented in terms of bond length differences for axial and equatorial bonds. The arbitrariness of this assumption turns out to be unimportant in the end as will be made clear. Using this table of coordinates and proceeding as described in chapter 2 for the tetrahedron, the point-charge representation of $V_{D_{4h}}$ may be shown to be:

$$\begin{aligned} V_{D_{4h}} = &-2(\sqrt{2\pi})\left(\sqrt{\frac{2}{5}}\right) ze\left(\frac{r^2}{a^3}-\frac{r^2}{b^3}\right) Y_2^0 \\ &+2(\sqrt{2\pi})\left(\sqrt{\frac{1}{72}}\right) ze\left(\frac{3r^4}{a^5}+\frac{4r^4}{b^5}\right) Y_4^0 \\ &+2(\sqrt{2\pi})\left(\sqrt{\frac{35}{144}}\right) ze\left(\frac{r^4}{a^5}\right)(Y_4^4+Y_4^{-4}). \end{aligned} \tag{5.9}$$

This may be rearranged to correspond to the form in (5.2):

$$\begin{aligned} V_{D_{4h}} = &-2(\sqrt{2\pi})\left(\sqrt{\frac{2}{5}}\right) ze\left(\frac{r^2}{a^3}-\frac{r^2}{b^3}\right) Y_2^0 \\ &-2(\sqrt{2\pi})\left(\sqrt{\frac{2}{9}}\right) ze\left(\frac{r^4}{a^5}-\frac{r^4}{b^5}\right) Y_4^0 \\ &+2(\sqrt{2\pi})\left(\sqrt{\frac{49}{72}}\right) ze\left(\frac{r^4}{a^5}\right)\left\{Y_4^0+\left(\sqrt{\frac{5}{14}}\right)(Y_4^4+Y_4^{-4})\right\}. \end{aligned} \tag{5.10}$$

We may note two features of this equation immediately. The first is that the third term in $V_{D_{4h}}$, i.e. V_{oct}, only involves the radial factor (r^4/a^5), which means that Dq is a function of equatorial bond lengths (a) but not axial ones (b). This repeats the point made above, after (5.4). The second feature is to note that the radial factors associated with Y_2^0 and Y_4^0 terms in the potential are similar and involve the functions $(r^k/a^{k+1}-r^k/b^{k+1})$. This is reflected in the matrix elements of the potential involving *differences* between quantities we have previously called radial parameters. Thus the coefficient of Y_2^0 leads to radial integrals of the form

$$\langle R(d)\left|\frac{r^2}{a^3}-\frac{r^2}{b^3}\right|R(d)\rangle = \langle R(d)\left|\frac{r^2}{a^3}\right|R(d)\rangle - \langle R(d)\left|\frac{r^2}{b^3}\right|R(d)\rangle. \tag{5.11}$$

Each term on the right-hand side of (5.11) is like Cp as defined in chapter 4 (4.12) except perhaps for a multiplicative factor. Similarly, associated with the coefficient of Y_4^0 there are radial integrals of fourth order:

$$\langle R(d)\left|\frac{r^4}{a^5}-\frac{r^4}{b^5}\right|R(d)\rangle = \langle R(d)\left|\frac{r^4}{a^5}\right|R(d)\rangle - \langle R(d)\left|\frac{r^4}{b^5}\right|R(d)\rangle, \tag{5.12}$$

each part of the right-hand side of (5.12) being equivalent to a Dq referred either to bond length a or b, respectively. By comparison of the definitions of Cp and Dq and of the matrix elements of $V_{D_{4h}}$ in (5.7) with those calculated explicitly with (5.10) we have the following relationships:

$$\left.\begin{aligned} Ds &= Cp(a) - Cp(b), \\ Dt &= \tfrac{4}{7}[Dq(a) - Dq(b)], \end{aligned}\right\} \tag{5.13}$$

where
$$Cp(a) = \tfrac{2}{7}ze^2\left(\frac{\overline{r^2}}{a^3}\right), \quad Dq(a) = \tfrac{1}{6}ze^2\left(\frac{\overline{r^4}}{a^5}\right),$$
$$Cp(b) = \tfrac{2}{7}ze^2\left(\frac{\overline{r^2}}{b^3}\right), \quad Dq(b) = \tfrac{1}{6}ze^2\left(\frac{\overline{r^4}}{b^5}\right).$$

$Dq(a,b)$ and $Cp(a,b)$ are the fourth-order and second-order radial integrals for the equatorial and basal ligands (a,b). The parameter Dt is a measure of the difference between the Dq values of equatorial and axial ligands and Ds is equal to the difference between the Cp values of these same ligands. Notice that since neither definition of Dq or Cp separates the bond length from the effective charge on the ligand (ze), the differentiation of axial and equatorial sites by bond length rather than ligand charge is immaterial.

There have been those, however, who choose to separate effective bond length from effective nuclear charge by factorizing $\overline{r^2}$ and $\overline{r^4}$ out from all the above equations. If this is done, we get

$$\left.\begin{aligned} Ds &= \tfrac{2}{7}ze^2\overline{r^2}\left(\frac{1}{a^3} - \frac{1}{b^3}\right), \\ Dt &= \tfrac{2}{21}ze^2\overline{r^4}\left(\frac{1}{a^5} - \frac{1}{b^5}\right), \end{aligned}\right\} \tag{5.14}$$

in which $\overline{r^2}$ and $\overline{r^4}$ are interpreted as mean-square and mean-fourth-power radii of the metal orbitals. This procedure implies that κ must be positive as $(1/a^3 - 1/b^3)$ must have the same sign as $(1/a^5 - 1/b^5)$. We do not favour the separation of $\overline{r^k}$ from $1/(a,b)^{k+1}$, not only for this reason but also because of the philosophy of parameterization involved: we shall make this clear in chapter 7. Lever *et al.*[67] choose to separate r^k from $(a,b)^{k+1}$ in an interesting way. In order to include the possibility of negative κ values, they define radial parameters proportional to $\overline{r^k}[1/a^{k+1} - A/b^{k+1}]$ rather than (5.14). This is interpreted to imply anisotropic metal radii r^k. Mathematically, this formalism is equivalent to that we use here in that the factorization of r^k and $(a,b)^{k+1}$ is effectively undone. However, these authors prefer to ascribe a degree of significance to bond lengths (a,b) which we consider indefensible. Their scheme allows them to conclude at one point that 'if the tetragonality of the Hamiltonian is due only to the chemical differences in the axial and equatorial ligands and not to differences in bond lengths, then the two parameters (Dt, Ds) *must* be both negative or both positive.' The force in this statement stems from a too-literal adherence to the point-charge model. As we discuss in chapter 7, bond lengths, like ligand charges and metal Z_{eff}, should be regarded as *effective* quantities in view of the fundamental deficiences of the point-charge approach. We prefer to blur the line between the roles of metal and ligand in this field and to limit such partition to qualitative discussion of *trends*. The scheme of Lever *et al.* gains no mathematical convenience over the simpler approach we describe and carries with it the penalty of an unnecessarily misleading partition of variables.

Taking Cp and Dq as basic parameters, as in (5.13), the following example shows how both positive and negative values of κ may arise.

Consider an MX_4Y_2 complex in which $Dq(X) = 0.8Dq(Y)$ and $Cp/Dq = 3$ for both sites. Substitution into (5.13) gives $Dt = -0.2 \times \frac{4}{7} Dq(Y)$ and $Ds = -0.2 \times 3Dq(Y)$ i.e. $\kappa = +\frac{21}{4}$. If the ratio Cp/Dq were changed to 4 for the equatorial (X) sites only, then we find $Dt = -0.2 \times \frac{4}{7}Dq(Y)$ as before, but $Ds = +0.2Dq(Y)$, giving $\kappa = -\frac{7}{4}$. It is not difficult to see how, in general, negative κ values can, but need not, derive from Cp/Dq ratios being different for basal and axial ligands but only for larger Cp/Dq ratios being associated with smaller Dq values. Thus writing

$$\left.\begin{aligned} r_1 &= Dq(b)/Dq(a), \\ r_a &= Cp(a)/Dq(a), \\ r_b &= Cp(b)/Dq(b), \end{aligned}\right\} \tag{5.15}$$

and substituting into (5.13) we find

$$\kappa = 7(r_a - r_b r_1)/4(1-r_1). \tag{5.16}$$

For κ to be negative r_b must increase as r_1 decreases or *vice versa*; explicitly, for κ to be negative,

$$r_b r_1 < \tfrac{4}{7}(r_1 - 1) + r_a. \tag{5.17}$$

It is the relationships in (5.13) and the little calculation above which suggest to us that Dq and Cp might be more significant, or at least more useful, parameters than Ds and Dt, because of the way parameters defined as *differences* can disguise many simple relationships. There is, however, the point that Dq, Dt and Ds form a set of three parameters as required by symmetry, whereas $Dq(a)$, $Dq(b)$, $Cp(a)$, $Cp(b)$ [or equivalently $Dq(b)$, r_1, r_a, r_b] form a set of four parameters and hence an extra and redundant degree of freedom. Nevertheless, it is possible to express *Dq*, *Dt* and *Ds* sets of experimental information more in the spirit of the latter set with interesting results. We do this now for some selected examples.

TETRAGONAL NICKEL(II) COMPLEXES

The term splitting diagram for spin-triplets in O_h and D_{4h} symmetry is as shown in figure 5.2. We quote, without proof, some simple theoretical facts about the energies of some of these levels.[c] The splitting $^3B_{2g} \leftrightarrow {}^3B_{1g}$ which normally yields the first or second spectral transition (depending on the sign of splitting of the $^3T_{2g}$ term) is given rigorously (but neglecting spin–orbit coupling) as $10Dq$ (equatorial). The first-order splittings of the $^3T_{2g}$ term is $\frac{35}{4}Dt$; of the $^3T_{1g}(F)$ term it is $-6Ds+\frac{5}{4}Dt$; and of the $^3T_{1g}(P)$ term, $3Ds-5Dt$. These last three quantities represent differences between the appropriate diagonal elements of the crystal-field potential, $V_{D_{4h}}$, and so are not necessarily very reliable measures of these quantities. Wherever possible we shall refer to data which have been obtained from complete diagonalization of the spin-triplet manifolds rather than direct use of the above formulae. Even so the role of spin–orbit coupling may cast doubt on such values in cases where this effect was neglected. This need not be a trivial matter as shown,[38] for example, by a study of the crystal spectrum and magnetic anisotropy of dichlorotetrakis(thiourea) nickel(II), $Nitu_4Cl_2$.

The first compounds we consider are *trans*-$Nipy_4Cl_2$ and $Nipy_4Br_2$. Ligand-field data for these complexes have been reported[93] as:

	Dt	*Ds*
$Nipy_4Cl_2$	286	525 cm^{-1}
$Nipy_4Br_2$	316	636

with *Dq*(py) in both cases close to 1100 cm^{-1}. Using the relationship

$$Dt = \tfrac{4}{7}[Dq(\mathrm{eq}) - Dq(\mathrm{ax})], \qquad (5.18)$$

we calculate *Dq* values for Cl and Br in these molecules as 600 and 550 cm^{-1}, respectively. Now let us represent the Cp/Dq ratios for Cl, Br and py as x, y and z, respectively. Using the relation

$$Ds = Cp(\mathrm{eq}) - Cp(\mathrm{ax}), \qquad (5.19)$$

we have: $$525 = 1100z - 600x, \qquad (5.20)$$

and $$636 = 1100z - 550y, \qquad (5.21)$$

whence $$y = 1.09x - 0.2. \qquad (5.22)$$

Also, from (5.20) $$z = 0.66x + 0.48, \qquad (5.23)$$

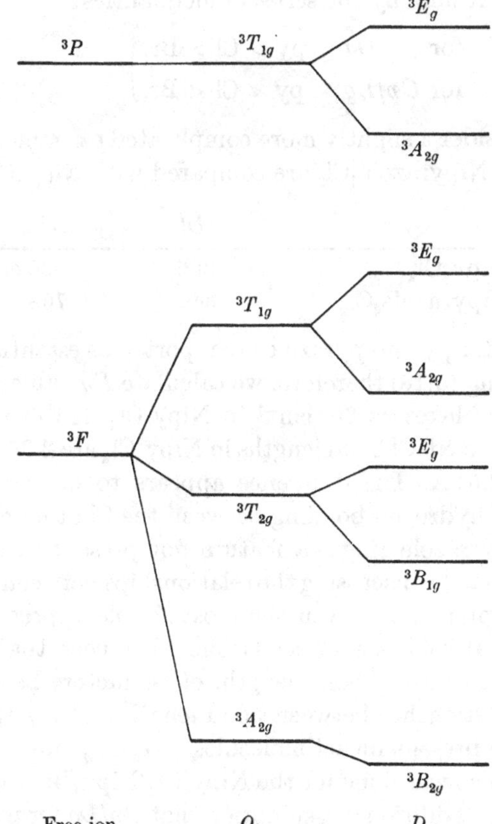

Figure 5.2. Splitting diagram for d^2 and d^8 systems in D_{4h} symmetry.

and hence the Cp/Dq ratios for Br and py may be expressed in terms of that for Cl. We can therefore construct the following table. *Assuming* Cp/Dq for Cl is greater than 2 we find that the Cp/Dq ratios for Br and py are respectively greater or less than that for chlorine. We might

Assumed Cp/Dq for Cl	Calculated Cp/Dq values for Br	py
2.0	2.0	1.6
3.0	3.1	2.1
4.0	4.2	2.7
5.0	5.3	3.2
	etc.	

summarize the results by the series of inequalities:

$$\left.\begin{array}{ll}\text{for} \quad Dq: & \text{py} > \text{Cl} > \text{Br},\\ \text{for } Cp/Dq: & \text{py} < \text{Cl} < \text{Br}.\end{array}\right\} \qquad (5.24)$$

We now consider a slightly more complicated example. The ligand-field data[91] for Nipyrazole$_4$Cl$_2$ are compared with Nipy$_4$Cl$_2$:

	Dt	*Ds*
Nipy$_4$Cl$_2$	286	525 cm^{-1}
Nipyrazole$_4$Cl$_2$	340	708

The *Dq* values for py and pyrazole are reported as essentially equal at 1100 cm^{-1}. Using (5.18) therefore, we calculate *Dq* values for chlorine in the two complexes as 600 cm^{-1} in Nipy$_4$Cl$_2$ and 505 cm^{-1} in Nipyrazole$_4$Cl$_2$. The Ni–Cl bond lengths in Nipy$_4$Cl$_2$ are 2.35 Å and in Nipyrazole$_4$Cl$_2$, 2.51 Å. The difference appears to be associated with intramolecular hydrogen bonding between the Cl atom and the N–H group of the pyrazole rings – a feature not present in the pyridine complex. We shall be discussing the relationships between bond length and *Dq* in chapter 7, but even the most simple appreciation of the origin of crystal-field theory leads one to expect that *Dq* should decrease with increasing bond length, other factors being constant. This simple relationship between bond length and *Dq* appears to be the case in the present nickel molecules. We may now proceed with our calculation as was done for the Nipy$_4$Cl$_2$/Nipy$_4$Br$_2$ case, provided we can make the additional assumption that *Cp/Dq* for py is the same as that for pyrazole. We hope this is a reasonable assumption on the basis of their similar *Dq* values.

Assumed *Cp/Dq* for Cl in Nipy$_4$Cl$_2$	Calculated *Cp/Dq* values for Cl in Nipyrazole$_4$Cl	Calculated *Cp/Dq* values for py/pyrazole
2.0	2.0	1.6
3.0	3.2	2.1
4.0	4.4	2.7
5.0	5.6	3.2

Once again these data may be summarized by the inequalities:

$$\begin{array}{ll}\text{for} \quad Dq: & \text{py/pyrazole} > \text{Cl(py)} > \text{Cl(pyrazole)},\\ \text{for } Cp/Dq: & \text{py/pyrazole} < \text{Cl(py)} < \text{Cl(pyrazole)},\end{array}$$

if *Cp/Dq* for Cl here is $\gtrsim 2$,

and so there is qualitative agreement between the data for the three complexes $Nipy_4Cl_2$, $Nipy_4Br_2$ and $Nipyrazole_4Cl_2$. Let us now consider compounds of a different metal.

PSEUDO-TETRAGONAL CHROMIUM(III) COMPLEXES

The paucity of low-symmetry ligand-field data means that we must illustrate our discussion with systems which are less than ideal. A pretty series is provided by the ions $Cr(en)_2X_2^{n+}$ where the *trans*-ligands X are H_2O, F, Cl and Br. The bidentate ethylenediamine ligands, however, reduce the molecular symmetry to D_{2h} at best, but we shall regard these molecules as pseudo-tetragonal for the present discussion. The splitting diagram for chromium(III) in D_{4h} symmetry is exactly the same as for nickel(II) in figure 5.2, except that all levels are spin-quartets. The same relationships between the various crystal-field matrix elements which were discussed for the nickel system, hold for d^3 systems. Ligand-field parameters for this series have been obtained from solution spectra[1,c] for the aquo-, chloro- and bromo-complexes and from crystal spectra[20] for the fluoro-derivative. They are as follows:

X	H_2O	F	Cl	Br
Dt	318	320	492	565 cm^{-1}
Ds	−302	−800	−291	−252
Dq(X) = *Dq*(axial)	1750	1600	1320	1190

The *Dq*(X) values were calculated using (5.18) and the observed *Dq* value for en of 2170 cm^{-1}. The latter was observed to be essentially independent of the nature of the axial groups.

It is also worth noting at this point that the calculated *Dq*(X) values agree well with those predicted by the Spectrochemical Series unlike those reported for the nickel(II) compounds earlier. This may be correlated with a much smaller degree of steric hindrance by the equatorial groups, or intramolecular hydrogen-bonding effects as in $Nipyrazole_4Cl_2$, in these chromium complexes.

Assumed *Cp*/*Dq* value for Cl	Calculated *Cp*/*Dq* value for Br	F	H_2O	en
2.0	2.2	2.0	1.5	1.1
3.0	3.3	2.8	2.3	1.7
4.0	4.4	3.6	3.0	2.3
5.0	5.5	4.5	3.8	2.9

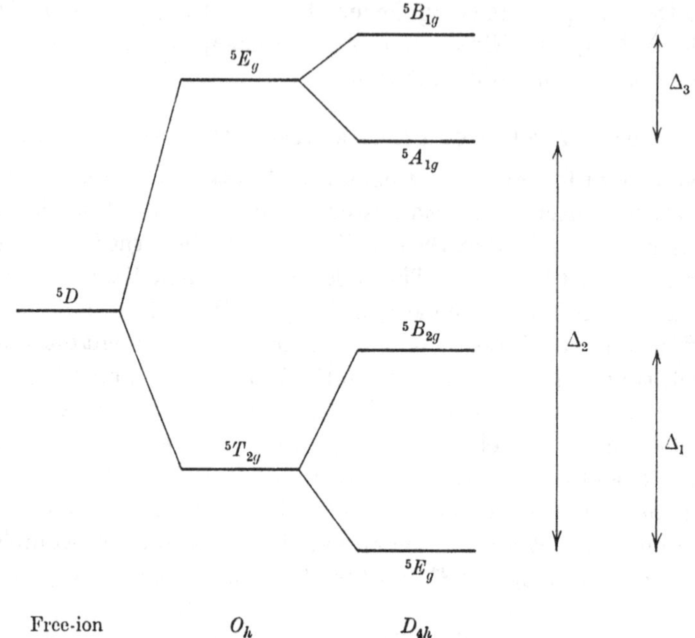

Figure 5.3. Splitting diagram and transition energies for d^4 and d^6 ions in D_{4h} symmetry.

A similar procedure to that used for the nickel molecules, for $Cp/Dq \gtrsim 2$, gives the above table, so that we have:

$$Dq: \quad \text{en} > H_2O > F > Cl > Br,$$
$$Cp/Dq: \quad \text{en} < H_2O < F < Cl < Br,$$

once more indicating the trends for Dq and Cp/Dq are opposed.

We turn now to examples for iron(II) for which there appear to be complications.

TETRAGONAL OCTAHEDRAL IRON(II) MOLECULES

The term splitting diagram for d^6 ions in D_{4h} symmetry is shown in figure 5.3. There are no difficulties associated with configurational interaction here, as there are no repeated term designations, so that term splittings may be expressed rigorously (spin–orbit coupling ignored) as differences of appropriate diagonal crystal-field matrix elements. It can be shown,[c] in terms of the definitions implied in figure 5.3, that:

$$\left.\begin{aligned} \Delta_1 &= 3Ds - 5Dt, \\ \Delta_2 &= 10Dq(\text{eq}) - Ds - 10Dt, \\ \Delta_3 &= 4Ds + 5Dt. \end{aligned}\right\} \qquad (5.25)$$

An alternative quadrate parameterization scheme is that of McClure[70] in which the *orbital* splittings are functions of $d\sigma$ for the octahedral e_g set and $d\pi$ for the t_{2g} set. This is supposed to have the advantage of separating σ- and π-bonding effects. We shall discuss the scheme in a later chapter. For present purposes it suffices to state the relationships between McClure's parameters and Dt and Ds:

$$\left.\begin{aligned} \Delta_3 &= -\tfrac{8}{3}d\sigma, \\ d\sigma &= -\tfrac{12}{8}Ds - \tfrac{15}{8}Dt. \end{aligned}\right\} \qquad (5.26)$$

Generally the splitting Δ_1 is too small to be observed directly *via* electronic absorption spectroscopy so that we usually are left with $d\sigma$ as the only directly observed low-symmetry parameter in d^6 systems. It is particularly useful when Δ_1 can be measured indirectly as a later example will show. Without this information we may only make progress in the manner of the nickel and chromium complexes already described by making further assumptions. We illustrate this by the case[10] of $Fepy_4Cl_2$ and $Fepy_4Br_2$ for which the observed $d\sigma$ values are -656 and $-1088\ cm^{-1}$, respectively. The Dq(py) value in these compounds is measured to be $1250\ cm^{-1}$ in either molecule. Let us assume Dq values for Cl and Br from the Spectrochemical Series as 800 and $760\ cm^{-1}$, respectively. Using (5.18) to calculate Dt values and then (5.26) for Ds, we may construct the following table:

	Cl	Br
$d\sigma$	-656	$-1088\ cm^{-1}$
Dt	257	280
Ds	115	376

From this we obtain, in the usual way:

$$\left.\begin{aligned} Dq&: \quad \text{py} > \text{Cl} > \text{Br}, \\ Cp/Dq&: \quad \text{py} < \text{Br} \lesssim \text{Cl}. \end{aligned}\right\} \qquad (5.27)$$

Thus, the reversed order of Cp/Dq with respect to Dq obtained so far is not observed here.

Another example[49] is afforded by the series $Fe(isoquinoline)_4X_2$ where X = Cl, Br, I. We have the following data:

X	Cl	Br	I	
$d\sigma$	-700	-1350	$-2200\ cm^{-1}$	
Dt	236	302	342	(5.28)
Ds	174	523	1046	

The $d\sigma$ values were deduced directly from spectra. Dt and Ds values were calculated in the same manner as for $Fepy_2X_2$ above taking an essentially constant value of 1130 cm^{-1} for Dq(isoquin) and 'Spectrochemical' Dq values for Cl, Br and I, as 680, 600 and 560 cm^{-1} respectively. These figures yield the series:

$$Dq: \quad \text{isoquinoline} > \text{Cl} > \text{Br} > \text{I},$$

$$Cp/Dq: \quad \text{isoquinoline} > \text{Br} > \text{Cl} > \text{I},$$

and, once again, the simple inversion of the series is not obtained.

The problem is rationalized, however, on re-examination of the $Fe(isoquinoline)_4X_2$ series in the light of Mössbauer studies[10] used to obtain values for the ground term splittings Δ_1 (figure 5.3). Although these Δ_1 values may be somewhat in error due to assumptions regarding the neglect of the excited terms from $^5E_g(O_h)$ and a constant and apparently arbitrarily chosen value for the orbital reduction factor, it is likely that the signs and orders of magnitude reported are reliable. Using the experimental values of Δ_2 and Δ_3 from the electronic absorption spectra and of Δ_1 from the Mössbauer work, Goodgame *et al.*[10] deduced Dt and Ds for the isoquinoline complexes from (5.25) as follows:

X	Cl	Br	I	
Dt	230	380	510 cm^{-1}	(5.29)
Ds	180	470	820	

These values differ quite markedly from those deduced, somewhat by guesswork, in (5.27). In particular, the Dq (axial) values implied by these Dt values and $Dq(\text{isoquinoline}) = 1130\,cm^{-1}$ are 695, 520 and 270 cm^{-1} for Cl, Br and I, respectively. We shall comment on these figures shortly.

Meanwhile, using the values in (5.29) and assuming $Cp/Dq\ (\text{Cl}) \gtrsim 2$, we obtain:

Assumed Cp/Dq value for Cl	Calculated Cp/Dq values for		
	Br	I	isoquinoline
2.0	2.2	2.8	1.4
3.0	3.5	5.4	2.0
4.0	4.9	8.0	2.6
5.0	6.2	10.6	3.3

and hence the sequences:

$$Dq\text{:}\quad \text{isoquinoline} > \text{Cl} > \text{Br} > \text{I},$$

$$Cp/Dq\text{:}\quad \text{isoquinoline} < \text{Cl} < \text{Br} < \text{I},$$

with the familiar inverse properties of Dq and Cp/Dq.

Very similar results and the same sequencing of Dq and Cp/Dq are found, for the series $Fe(\gamma\text{-picoline})_4X_2$ (X = Cl, Br, I) using Goodgame's $\Delta_1 \Delta_2$ and Δ_3 values obtained from electronic and Mössbauer spectra.[10]

If, by comparison with the $Fe(\text{isoquinoline})_4X_2$ series, we reassess Dq(X) values for the Cl and Br ligands in $Fepy_4X_2$ as 700 and 550 cm^{-1}, respectively, we again obtain the ordering:

$$Dq\text{:}\quad \text{py} > \text{Cl} > \text{Br},$$

$$Cp/Dq\text{:}\quad \text{py} < \text{Cl} < \text{Br}.$$

Dq VALUES: FURTHER EXAMPLES

It appears, therefore, that the reversed ordering of Cp/Dq compared with Dq may be a common feature of tetragonal molecules but, in order to reproduce the sequence in some iron(II) complexes, one must accept the very low Dq values quoted above. In particular, consider the Dq value for I in $Fe(\text{isoquinoline})_4I_2$ of 270 cm^{-1}. Comparison with the Spectrochemical Series suggests this value is a factor of two too low. But it would require a Δ_1 value of *ca.* 1200 cm^{-1} rather than the -80 cm^{-1}, observed, to reproduce the 'Spectrochemical' Dq value for I as *ca.* 800 cm^{-1}. We feel sure the reported Dq value of 270 cm^{-1} is substantially correct and we further believe that it reflects a considerable degree of steric hindrance in the molecule. The situation throughout all the tetragonal molecules discussed above is probably similar to that in $Nipy_4Cl_2$ and $Nipyrazole_4Cl_2$. The pyridine, pyrazole, ethylenediamine, isoquinoline and γ-picoline ligands pack around the central metal atom in such a way as to reduce in-plane inter-ligand repulsions but the more bulky of these inevitably provide steric hindrance to the axial halogen, or other, ligands. This would be particularly important for the larger axial ligands like Br^- and I^- which, on being repelled further from the metal, are associated with 'anomolously' low Dq values. Hydrogen-bonding effects, as in the pyrazole complex, may be collected under the umbrella description of 'steric hindrance' for these purposes. As the Dq(axial) values for the $Cr(en)_2X_2$ series, deduced from observed Dq(en) and Dt values, agree well with

'Spectrochemical' values, we may deduce that steric hindrance between the equatorial ethylenediamine groups and the axial ligands is minimal, despite the possibilities offered by the puckering involved in the diamine chain. Dq(Cl) values in $Co(NH_3)_4Cl_2^+$ and $Co(en)_2Cl_2^+$ have been reported[c] as 1258 and 1138 cm^{-1}, respectively, which may indicate slight steric interaction in the ethylenediamine complexes, but such small differences might merely reflect the bidentate nature of the en ligands and loss of D_{4h} symmetry. The situation is clearer[c] in the ions, Ni(*N*, *N*-dimethyl-en)$_2$(H_2O)$_2^{2+}$ compared with $Ni(en)_2(H_2O)_2^{2+}$, for which $Dq(H_2O)$ values are 459 and 590 cm^{-1}, respectively, reflecting the more bulky nature of the substituted ethylenediamine ligand.

Another series[38, 43, 49] which appears to illustrate the effects of intramolecular hydrogen bonding is *trans*-$M^{II}tu_4Cl_2$ where tu = thiourea and M = Ni, Co, Fe, Mn. Recent studies of magnetic anisotropies and single-crystal spectra have shown that the equatorial thiourea ligands and the axial chlorines have nearly equal Dq values in the cobalt, iron and manganese complexes. Their relative size, i.e. the sign of Dt, could not be established but both Ds and Dt have the same sign and are small. The situation is different in the nickel analogue, however, in which Dq(thiourea) = 930 and Dq(Cl) $\sim$ 600 cm^{-1}. The crystal and molecular structure of both nickel and cobalt analogues, with which latter the iron and manganese compounds are isomorphous, have been determined by X-ray crystallography. While all compounds in the series crystallize in tetragonal space groups the nickel molecules lack the centre of symmetry the cobalt ones possess. The Ni–Cl bond lengths are unequal, one chlorine atom being extensively hydrogen-bonded intramolecularly to the thiourea ligands, and other *via* intermolecular hydrogen-bonds. The average Ni–Cl bond length is anomolously large while the Co–Cl bond lengths are normal. These structural features appear to be faithfully reproduced in the ligand-field parameters deduced from the single-crystal studies so, incidentally, lending support to the theoretical models employed. The Ds value in $Nitu_4Cl_2$ is $\leqslant 350$ cm^{-1} and the usual inverse relationship between Dq and Cp/Dq again follows. This study also emphasized the role of spin–orbit coupling in these calculations. The Ds value was estimated from the splitting of the $^3T_{1g}(F)$ octahedral term. A large fraction of this splitting derives from the effects of spin–orbit coupling:[38] had this been neglected, the estimated Ds value would have been *ca.* 1000 cm^{-1} instead of $\leqslant 350$ cm^{-1}.

Finally we consider an example of some five-coordinate molecules.

Caride *et al.*[11] have examined the spectra of the square-pyramidal cobalt complex ions $Co(MeCN)_5^{2+}$ and $Co(PhCN)_5^{2+}$. The essentially C_{4v}-symmetry crystal field was parameterized in a similar way to that discussed for D_{4h} and the following parameter values were obtained:

	$Co(MeCN)_5^{2+}$	$Co(PhCN)_5^{2+}$
Dq(basal)	4230	4230 cm^{-1}
Ds	−2350	−2060
Dt	−2710	−2640

Axial *Dq* values calculated from these data are:

Dq(axial)	8960	8860 cm^{-1}

Assuming similar *Cp*/*Dq* ratios for *basal* MeCN and PhCN ligands (which have the same *Dq* values) we construct the following table:

Assumed *Cp*/*Dq* value for PhCN (axial)	Calculated *Cp*/*Dq* values for MeCN (axial)	basal ligands
1.0	1.02	1.6
2.0	2.01	3.7
4.0	3.99	7.9
6.0	5.97	12.1

and so find the same relationship already seen for D_{4h} molecules, *viz.*

$$Dq: \quad \text{MeCN(axial)} \gtrsim \text{PhCN(axial)} > \text{basal ligands},$$

$$Cp/Dq: \quad \text{MeCN(axial)} \lesssim \text{PhCN(axial)} < \text{basal ligands}.$$

It is worth pointing out here that it is quite difficult to measure values of low-symmetry ligand-field parameters, single-crystal studies usually being required. Even with the extra information contained in polarized spectra, magnetic anisotropies and the like, parameter values are frequently only obtainable approximately, because the sophistication of the techniques (as compared with powder and solution work) requires proper treatment of spin–orbit coupling, orbital reduction effects and so on. Ambiguities in fitting experimental data to theory are common and these are often intrinsic rather than reflections of inadequacies in data or theoretical models. At this stage, therefore, we consider it best to discuss these parameter values in a semi-quantitative fashion and not rely too heavily on the exact numbers frequently quoted.

SUMMARY

Before moving on to a theoretical discussion of ligand-field radial parameters it is appropriate to re-examine the computational origin of the Dq:Cp/Dq relationship we have demonstrated. The process of deducing Cp/Dq ratios from Dt and Ds data was exemplified by (5.20) to (5.22), shown explicitly for the $Nipy_4X_2$ system. Writing (5.22) in terms of the more explicit symbols Dq and Cp, we have:

$$Cp/Dq(\mathrm{Br}) = \frac{Dq(\mathrm{Cl})}{Dq(\mathrm{Br})} \cdot Cp/Dq(\mathrm{Cl}) - \frac{\{Ds[\mathrm{Br\ compound}] - Ds[\mathrm{Cl\ compound}]\}}{Dq(\mathrm{Br})}. \tag{5.30}$$

Clearly the inverse relationship between Cp/Dq and Dq would follow precisely if the difference between the Ds values for the two compounds vanished, for then:

$$[Dq \times (Cp/Dq)]\,(\mathrm{Br}) = [Dq \times (Cp/Dq)]\,(\mathrm{Cl}). \tag{5.31}$$

Thus, the best chance of the inverse relationship breaking down is for disparate Ds values of the two compounds involved in (5.30). Further such significant differences in Ds values would have to be in one sense only. Thus in (5.30) a breakdown in the relationship would require Ds(Br complex) > Ds(Cl complex) rather than the other way round. Also such a breakdown in the inverse relationship would occur more easily for overall low Cp/Dq ratios than for high, reflecting the relative dominance of the two terms on the right-hand side of (5.30).

Frequently the experimental data do give Ds increasing with Dq of the axial ligand, in the present context, as shown in the values quoted for $Fepy_4X_2$ and $Nipy_4X_2$ for example but the differences between the appropriate Ds values are small in comparison with Dq [the denominator of the second term in (5.30)]. It is probably the small absolute magnitude of Ds values as much as anything which is ultimately responsible for establishing the inverse relationship between Dq and Cp/Dq: or, of course, *vice versa*.

One thing that can be deduced from this little discussion is that the inverse relationship is probably genuine, as whatever errors there are in the raw experimental data surely cannot be so large as to underestimate Ds values by a sufficient amount (5 × ?) to reverse the trend. For example, the Ds value for $Cr(en)_2F_2$ reported above seems out of line with those for the aquo-, chloro- and bromo-ions, this no doubt being the result of a single-crystal study versus solution studies. The disparity does not, in any case, upset the inverse relationship between *Dq* and *Cp*/*Dq*, *provided that Cp*/*Dq ratios are not too small.*

In this and the preceding chapter we have defined some low-symmetry crystal-field parameters, demonstrated their relevance to experimental observables and acquired a feeling for their relative and, sometimes, absolute magnitudes. *Cp* appears to be larger than *Dq*, perhaps much larger, and the ratio *Cp*/*Dq* tends to increase with coordination number. Study of the quadrate parameters *Ds* and *Dt* may suggest inverse trends in the behaviour of *Cp*/*Dq* relative to *Dq*. This same trend (based on two compounds!) was also deduced from the magnetic properties of trigonally-distorted octahedral cobalt(II) complexes, as expressed in (4.18). Thus, the data, as far as it goes, does show a consistent and discernable pattern. It is obviously desirable to check this pattern with many more studies of the sort discussed. It is now of great interest to see if this pattern makes sense in terms of the point-charge model we have made so much use of: in particular, whether these trends correlate with more general chemical concepts like the electroneutrality principle, electronegativity and so on. We examine this in chapter 7, pausing first to examine the nature of 10*Dq* as revealed in calculations of widely ranging sophistication.

GENERAL REFERENCES

(*a*) C. J. Ballhausen, ***Introduction to Ligand Field Theory***, McGraw-Hill, New York, 1962.

(*b*) B. N. Figgis, ***Introduction to Ligand Fields***, Interscience, New York, 1966.

(*c*) A. B. P. Lever, The Electronic Spectra of Tetragonal Metal Complexes: Analysis and Significance, ***Coord. Chem. Rev.*** 1968, **3**, 119.

6
THE NATURE AND CALCULATION OF 10Dq

We have seen how the apparent success of crystal-field theory depends partly upon symmetry and partly on parameterization of the strength of the crystal field. In this same spirit of empiricism, the equivalence of the crystal-field and the qualitative molecular-orbital theories is apparent, although the interactions between the metal ion and its surrounding ligands are different in the two theories. In this chapter we are concerned with the quantitative aspects of these theories and particularly with their abilities to account for the d orbital splittings as epitomized by 10Dq in octahedral complexes. We are not concerned with the methodology of computations; nor with topics such as the transferred hyperfine effects and neutron form-factors, both of which, being as much the result of metal–ligand interactions as is 10Dq, have been used to check the results of such calculations. The various estimates of 10Dq will be considered essentially from an historical point of view because this allows us to appreciate how the various theoretical shortcomings have led to modifications of the models employed. In turn, there emerges from these modifications a reappraisal of our understanding of the physical nature of metal–ligand interactions and of the various factors contributing to 10Dq. To a much lesser extent we consider how the theoretical inadequacies are complemented and reinforced by experimental data since the accurate prediction of, or agreement with, such data remains the primary objective of any *ab initio* calculation.

The MO calculations should be seen as attempts to calculate spectroscopic transition energies rather than one-electron radial integrals. The identification of the ${}^3A_{2g}-{}^3T_{2g}$ splitting in octahedral nickel(II) ions, for example, with 10Dq begs important questions of definition. While these are discussed at the end of this chapter, we follow the various authors whose work we review by referring to their efforts as calculations of 10Dq, *pro tem.*

THE ELECTROSTATIC OR IONIC CRYSTAL-FIELD THEORY

Bethe's theory concerning the origins of term splittings in ionic crystals provided the first physical model upon which calculations of $10Dq$ were based. Thus, Van Vleck[113] calculated a value of $10Dq$ for the $Cr(H_2O)_6^{3+}$ complex ion in chrome alum using a point-dipole representation of the water molecule and a hydrogenic function for the $3d$ orbital of the chromium ion. This was done essentially by evaluation of the integral in (2.66). Good agreement was obtained between calculated and observed $10Dq$ values although Van Vleck was careful to point out that such agreement was possibly fortuitous in view of the crude approximations inherent in this model. Further estimates of $10Dq$ were made by Polder[85] using a point-charge model and again, reasonable agreement with experiment was obtained. The results of calculations such as these were generally taken as evidence for the essential correctness of the electrostatic crystal-field theory.

This attitude persisted until the 1950s at which time two important developments occurred. The first of these concerned the rapidly increasing amount of experimental evidence, mainly from magnetic resonance techniques, indicating the importance of electron transfer and covalency in metal complexes. Although covalent bonding could be incorporated into the mathematical framework of the electrostatic theory (e.g. by scaling factors like Stevens' orbital reduction parameter[105, 41]), it became clear that some of the basic assumptions of the theory needed to be re-examined. The second development of interest concerns the availability of better atomic wave-functions.

In an attempt to remedy some of the theoretical and conceptual weaknesses of the electrostatic theory, Kleiner[62] calculated Dq, again for chrome alum, using a more realistic charge distribution for the ligands. In this model, there was a contribution to Dq from the assumed dipolar water ligands as in Van Vleck's earlier calculation, but a further contribution from the Coulomb interaction of the $3d$ electrons with the electron cloud of the oxygen atoms came from two-electron interelectron repulsion terms. A third contribution to Dq derived from the electrostatic interaction of the metal electrons with the positively charged oxygen nuclei – a penetration effect. The classical dipolar contribution to Dq in $Cr(H_2O)_6^{3+}$ was found to be *ca.* $1000\ cm^{-1}$, whereas a negative contribution from the 'correction' terms was calculated at *ca.* $-1500\ cm^{-1}$. The best value of Dq was thus found to be $-500\ cm^{-1}$ indicating that the dominant electrostatic interaction of the d electrons was with the ligand nuclei: further, this 'best value' of Dq required a dipole moment for the water ligand three times that of the free water ligand, consistent, it was supposed, with a polarization of the ligand charge density by the metal ion.

A similar model was employed by Tanabe and Sugano[108] who included the exchange interactions between the metal and ligand electrons in addition to the Coulomb terms above. Also, metal and ligand orbitals were orthogonalized so that the wavefunctions used consisted of metal $3d$ orbitals with admixed ligand $2s$ and $2p$ functions. Although this was a considerable departure from the crude electrostatic picture it did not represent a true molecular-orbital picture. Kleiner's value for Dq was improved but although the sign was corrected, the magnitude was overestimated by a factor of two.

Kleiner's calculation was later re-examined by Freeman and Watson[30] using Watson's Hartree–Fock atomic wavefunctions.[115] These SCF functions are radially much contracted relative to the Slater functions used in previous calculations (see later) and as a result, much smaller magnitudes of both the dipolar and Coulomb correction terms were calculated. Although the negative sign of Dq obtained by Kleiner was reversed, the best value found was only a few hundred wavenumbers. In the same paper, Freeman and Watson showed that the inclusion of orthogonalization of metal and ligand orbitals did not cancel Kleiner's correction term, thus refuting a suggestion by Phillips[83] that all terms arising from the extension of the electron clouds of the ligands cancelled to leave the dipolar term as the sole contribution to Dq. This latter claim had led Phillips to calculate a value of *ca.* 1200 cm^{-1} for Dq in chrome alum.

These calculations indicate how increasing 'correctness' regarding the atomic wavefunctions has led to a worsening of the calculated Dq values. We can understand this trend by considering the different shapes of the orbitals used for these various calculations,[104] shown schematically in figure 6.1.

The figure shows how the hydrogenic functions used by Van Vleck and by Polder are much expanded with respect to the SCF functions. Kleiner used Hartree functions which were calculated without inclusion of exchange integrals. As discussed in chapter 3, exchange gives rise to electron–electron attraction effects and this is demonstrated by the correspondingly more contracted Hartree–Fock functions used by Freeman and Watson which include exchange terms. Thus we see that the earliest calculations of Van Vleck and of Polder gave good values of Dq because the over-expanded metal orbitals gave an erroneously

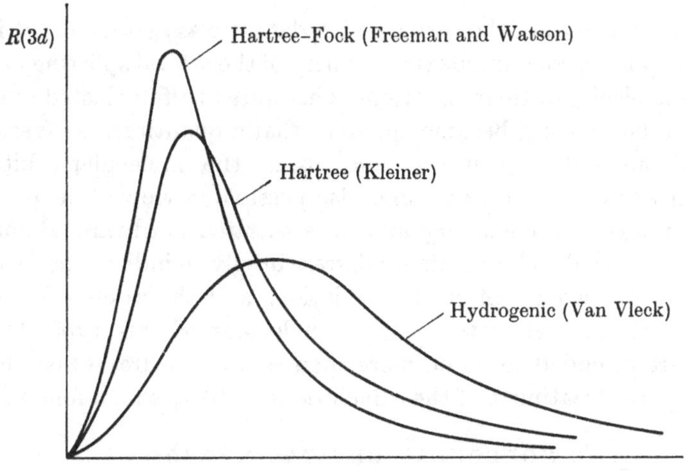

Figure 6.1. Radial *d* wavefunctions (schematic) used in various approximations.

large contribution, their electrostatic interaction with the ligands being thereby overestimated. Similarly, a comparison of the Hartree and Hartree–Fock orbitals shows how the Freeman and Watson calculations led to a reduction in both the dipolar and correction terms with respect to Kleiner's values, consistent with the decreased interaction of the metal electrons with both the ligand electrons and nuclei. Freeman and Watson concluded that the electrostatic crystal-field theory was incapable of predicting values of Dq. Refinements of the model aimed at improving its physical realism have led to further difficulties, not the least of which is that the theoretical splitting of the *d* orbitals is negligible! These theoretical failures, together with the ever-increasing weight of experimental evidence of the importance of covalency, has caused the electrostatic theory to be abandoned as a fundamental theory describing metal–ligand interactions. Good reasons remain, however, for using the electrostatic theory in an empirical way: these are described in the next chapter.

THE MOLECULAR-ORBITAL FORMALISM[a,b]

The bonding in transition-metal complexes has been considered within the molecular-orbital framework since 1935 but no quantitative investigations were made for many years, partly because of the inherent difficulty of such calculations until the advent of high-speed computers and also because the value of Dq had been successfully obtained within the simpler electrostatic theory. Even though covalency was evident from the earliest e.s.r. data it was assumed that in these mainly ionic compounds the majority of the orbital splitting had its source in electrostatic interactions. This state of affairs lasted until the early 1960s when it became apparent that a quantitative investigation of metal–ligand interactions using the molecular-orbital formalism was highly desirable and also possible in view of developments in theoretical chemistry and its associated hardware. Before considering these developments we digress briefly to indicate some of the general philosophy of such calculations and the application of molecular-orbital theory to complex molecules of this type. Our remarks are intended to do no more than establish a framework for the descriptive treatment of the calculations of 10*Dq* which follow.

THE MOLECULAR-ORBITAL METHOD

The molecular orbitals which describe the behaviour of each electron moving in a potential field due to the nuclei and the other electrons are eigenfunctions of a suitably defined one-electron Hamiltonian. The interaction between the *i*th electron and all other electrons may be described by an averaged potential field and represented by the equation:

$$\mathscr{F}\psi = \epsilon\psi, \tag{6.1}$$

where $\mathscr{F}$ is the Fock operator of the form:

$$\begin{aligned} \mathscr{F} &= \mathscr{H} + \mathscr{G} \\ &= \mathscr{H} + \sum_{i=1}^{n} (2J_i - K_i). \end{aligned} \tag{6.2}$$

$\mathscr{H}$ is the core Hamiltonian comprising the electron-nuclear attraction term and the kinetic energy operator and $\mathscr{G}$ describes the interaction between the *i*th electron and the other electrons. $\mathscr{H}$ and $\mathscr{G}$ are one- and two-electron properties, respectively. In the case of a closed-shell electronic configuration $\mathscr{G}$ is simply expressed as in (6.2) where J and K are Coulomb and exchange terms, these interactions being summed over the n doubly occupied orbitals. By definition J and K involve the orbitals ψ to be obtained in (6.1). Accordingly (6.1) must be solved by an iterative procedure. Thus for a zeroth-order set of orbitals, the operators are defined from which a new set of orbitals may be obtained by solution of (6.1): the process is repeated until the Hamiltonian obtained is consistent with the orbitals that it determines and that determine it. Then $\mathscr{F}$ is said to be a self-consistant field (SCF) Hamiltonian and is considered as giving the averaged potential field acting upon any one electron.

This procedure reviews the standard SCF approximation for solving the Schrödinger equation for a many-electron system and has been used extensively to obtain wavefunctions for many-electron atoms. The extension of the method to the realm of molecular calculations involves the introduction of other approximations which become more involved as the molecule becomes more complex. The most common such assumption employed in this area is the linear combination of atomic orbitals (LCAO) approximation. If we take a complete set (in principle, infinitely large) of basis atomic orbitals we could obtain exact SCF molecular orbitals by a suitable choice of the expansion coefficients c_i in (6.3):

$$\psi = \sum_i^{\infty} c_i \phi_i. \tag{6.3}$$

The molecular orbitals obtained with a sufficiently large basis set have energies close to the true Hartree–Fock limit, while for restricted or incomplete basis sets the best orbital energy is obtained by applying the variation principle. This procedure leads to the Roothaan equations for the c_i from which the orbital energies are obtained. We do not pursue this matter further but those interested will find the Roothaan–Hartree–Fock method amply documented elsewhere.[a, 74, 92]

However, we do introduce one pertinent aspect here. This concerns systems (atoms or molecules) with open-shell configurations, in which there is an excess of electrons of one spin relative to the other. In these circumstances the potential 'seen' by an up-spin electron will be different from that seen by a down-spin electron by virtue of the different exchange contributions from the other electrons of the same spin. Accordingly, the Fock operator will be different for the two sets of electrons and consequently the orbitals will have different spatial character and different energies for up- and down-spin electrons. Two approaches have been used in this connection. The Restricted Hartree–Fock (RHF) method seeks to average the exchange terms to obtain a single orbital energy. The Unrestricted Hartree–Fock (UHF) approach treats the 'spin-splitting' explicitly, describing different spatial functions for up- and down-spin electrons. These remarks are particularly relevant for transition-metal ions and complexes which frequently involve open-shell configurations and the calculated orbital energies and radial behaviour of these orbitals depend on the choice of RHF or UHF computational procedures.

Our superficial discussion of the SCF basis of orbital theory has not differentiated in principle between atoms and molecules. If it were not for the evaluation of the numerous integrals which occur in molecular calculations, the two cases would be virtually identical, for the subsequent application of the Roothaan equations presents no further problems in the molecular case. In practice, of course, the vastly increased level of complexity molecular calculations involve, derives almost entirely from the evaluation of these integrals. In order to establish some sense of perspective, it seems worth while to sketch the sorts of problems these calculations entail. This leads naturally to the important choice of basis functions.

The atomic orbitals ϕ in (6.3) are expressable as linear combinations of Slater-type orbitals (STOs) of the form:[a]

$$\phi(r,\zeta) = R_{nl}(r,\zeta) \, . \, Y_l^m, \tag{6.4}$$

where $R_{nl}(r,\zeta) = r^{n-1} . e^{-\zeta r}$ and is the radial part of the Slater orbital: Y_l^m represents the angular part as spherical harmonics. The STOs may be combined giving:

$$R(r) = \sum_{r,\zeta} c_i R_{nl}(r,\zeta), \tag{6.5}$$

and for a given set of STOs with specified zeta values, the Roothaan SCF equations minimize the energy with respect to the expansion coefficients c_i. The 'best' atomic wave-functions, involving four or five STOs per atomic orbital, are Watson's SCF functions but for many purposes a smaller number is convenient; examples[a] of these are Richardson's double-zeta functions which are optimized in such a way that one STO represents orbital behaviour close to the nucleus and the second STO the behaviour far from the nucleus. Another is functions using one STO per atomic orbital (single-zeta functions) which are still less accurate but give a semi-quantitative picture of orbital behaviour.

Molecular orbitals may be formed from these various atomic basis functions in several ways. The crudest is a linear combination of single-zeta functions centred on the various nuclei while greater accuracy has been obtained by taking combinations of double-zeta functions. However, because of the large number of difficult integrals to be evaluated in molecular calculations it has frequently been found convenient to employ Gaussian-type orbitals[a] (GTOs) in place of the STOs. In principle, this leads to great simplification in the evaluation of integrals since the product of two GTOs centred on different nuclei is equivalent to a single GTO at a new centre. However, the radial behaviour of GTOs is such as to require a considerably greater number of them relative to STOs in expansions equivalent to (6.5) and so the number of integrations to be performed is much increased. There arises the practical problem of choosing between STO and GTO bases, involving fewer difficult integrations or more, less difficult integrations. This is obviously a technical matter dependent more on computer characteristics than on any chemical aspect. An idea of the magnitude of the computational problem is given by the NiF_6^{4-} cluster with a total of 86 electrons. An SCF all-electron calculation approximating the Hartree–Fock limit would require some 10^8 integrations if GTOs were used. The use of restricted basis sets for many of the calculations shortly to be discussed should occasion little surprise.

COVALENCY

The molecular orbitals we require for discussion of $10Dq$ in octahedral complex ions are defined[c] as

$$\left.\begin{aligned} \psi_e^A &= N_e(\phi_e - \lambda_s \chi_{2s} - \lambda_\sigma \chi_{2p\sigma}), \\ \psi_t^A &= N_t(\phi_t - \lambda_\pi \chi_{2p\pi}), \end{aligned}\right\} \tag{6.6}$$

for the antibonding MOs; correspondingly, for the bonding orbitals:

$$\left.\begin{aligned} \psi_{es}^B &= N'_{es}(\chi_{2s} + \gamma_s \phi_e + \gamma_{s\sigma} \chi_{2p\sigma}), \\ \psi_{e\sigma}^B &= N'_{e\sigma}(\chi_{2p\sigma} + \gamma_\sigma \phi_e + \gamma_{\sigma s} \chi_{2s}), \\ \psi_t^B &= N'_t(\chi_{2p\pi} + \gamma_\pi \phi_t), \end{aligned}\right\} \tag{6.7}$$

where the ϕs are metal atomic orbitals and the χs, linear combinations of the ligand atomic orbitals constructed by standard group-theoretical procedures. The normalizing coefficients N are determined by the coefficients γ and λ and by the metal–ligand and, where appropriate, ligand–ligand, overlap integrals. The coefficients γ and λ are measures of the covalency in bonding and antibonding MOs, respectively. As the MOs of the same symmetry are orthogonal, these coefficients are not independent. Thus for orbitals of t_{2g} symmetry we require that

$$\int \psi_t^{A*} \psi_t^B \, d\tau = 0,$$

i.e.

$$\int (\phi - \lambda\chi)^* (\chi + \gamma\phi) \, d\tau = 0,$$

where we have dropped the subscripts t and $2p\pi$ for convenience. Hence:

$$\int \phi^* \chi \, d\tau + \gamma \int \phi^* \phi \, d\tau - \lambda \int \chi^* \chi \, d\tau - \lambda\gamma \int \chi^* \phi \, d\tau = 0,$$

so that

$$S + \gamma - \lambda - \lambda\gamma S = 0, \tag{6.8}$$

when the orbitals on the various atoms are normalized and S is the overlap integral between metal and the symmetry-adapted ligand function. Therefore:

$$\lambda = \frac{\gamma + S}{1 + S\gamma}, \tag{6.9}$$

or, if we consider only terms to first order in overlap and covalency, we have:

$$\lambda_i = \gamma_i + S_i, \tag{6.10}$$

for the ith MO given in (6.6) or (6.7).

It is unfortunate that the term *covalency* has a somewhat vague meaning when applied to chemical bonds. It is difficult to give a precise definition except to say that our future discussion rests on the use of the mixing coefficients γ (or λ) to measure the extent of covalency in metal–ligand bonds.

ONE-ELECTRON CALCULATIONS

The forerunner of *ab initio* calculations of $10Dq$ within the framework of molecular-orbital theory was that performed by Sugano and Shulman (SS)[97] for the NiF_6^{4-} octahedral cluster in $KNiF_3$. These authors obtained good agreement between their calculated value of $10Dq$, 6350 cm^{-1}, and the experimental value of 7250 cm^{-1}. The reductions of the spin–orbit coupling coefficient and of Racah's B parameter from their free-ion values were also satisfactorily explained. However, the model upon which the SS calculation was based was later shown to be invalid concerning the source of the observed covalency. This specific criticism was contained in the work of Watson and Freeman (WF)[c] who further presented a detailed calculation for the same cluster emphasizing in great detail the source and nature of the overlap and covalency effects that contribute to the crystal-field splitting. We now review the WF calculation to examine how it affects our understanding of the physical picture of the metal–ligand interaction.

Watson and Freeman's main criticism of the SS calculation centres around their choice of the antibonding electrons as the source of the observed covalency effects – crystal-field splitting and transferred hyperfine effects – thus requiring the evaluation of the λs in (6.6). However, for a given pair of orthonormalized molecular orbitals (a bonding orbital and its antibonding counterpart of the same spin and symmetry designation) the *covalency* of the antibonding electrons is exactly compensated by the covalency of the bonding partners. The

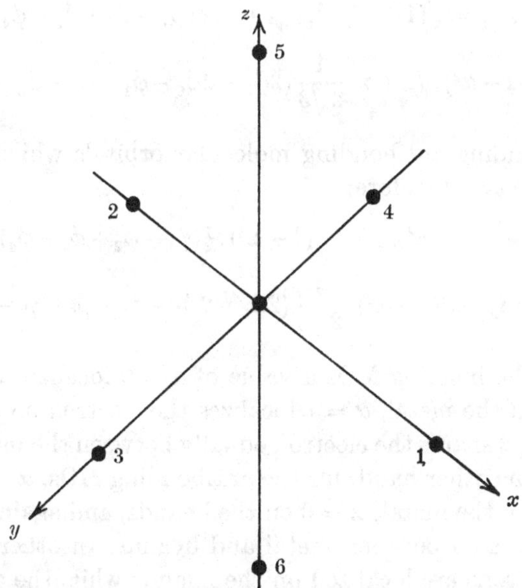

Figure 6.2. Numbering of ligands in the octahedron.

subtle distinction between the roles of the bonding and antibonding electrons concerning covalency and overlap effects is an important point in any discussion of the source of the crystal-field splitting. Often covalency in metal–ligand bonds is discussed in loose terms: thus, 'dilution of metal d orbitals', 'expansion of d electrons onto the ligands' are the sort of terms frequently encountered. By implication the covalency and its experimental manifestations are associated with the essentially d-like antibonding molecular orbitals. The failings of this simplistic view should be apparent after the following discussion.

Let us consider the e_g molecular orbitals and describe them by nomenclature more commonly used in chemistry. Using the numbering scheme and axes shown in figure 6.2, we may construct the following pair of molecular orbitals transforming as e_g in O_h symmetry:

$$e_g \begin{cases} \psi_{x^2-y^2} = \beta d_{x^2-y^2} + \alpha \,.\, \tfrac{1}{2}(\phi_1 - \phi_2 + \phi_3 - \phi_4), & (6.11) \\ \psi_{z^2} = \beta d_{z^2} + \alpha \,.\, \dfrac{1}{2\sqrt{3}}(2\phi_5 + 2\phi_6 - \phi_1 - \phi_2 - \phi_3 - \phi_4), & (6.12) \end{cases}$$

where the ϕs are ligand σ-type orbitals and α and β are the mixing coefficients. Ignoring overlap between metal and ligand orbitals, allows us to write:

$$\alpha^2 + \beta^2 = 1,$$

so that: $$\psi_{x^2-y^2} = \sqrt{(1-\alpha^2)}\,.\,d_{x^2-y^2} + \alpha\,.\,\tfrac{1}{2}(\phi_1 - \phi_2 + \phi_3 - \phi_4), \qquad (6.13)$$

$$\psi_{z^2} = \sqrt{(1-\alpha^2)}\,.\,d_{z^2} + \alpha\,.\,\frac{1}{2\sqrt{3}}(2\phi_5 + 2\phi_6 - \phi_1 - \phi_2 - \phi_3 - \phi_4). \quad (6.14)$$

The corresponding antibonding molecular orbitals which are orthogonal to these are therefore:

$$\psi^*_{x^2-y^2} = \alpha d_{x^2-y^2} - \sqrt{(1-\alpha^2)}\,.\,\tfrac{1}{2}(\phi_1 - \phi_2 + \phi_3 - \phi_4), \qquad (6.15)$$

$$\psi^*_{z^2} = \alpha d_{z^2} - \sqrt{(1-\alpha^2)}\,.\,\frac{1}{2\sqrt{3}}(2\phi_5 + 2\phi_6 - \phi_1 - \phi_2 - \phi_3 - \phi_4). \quad (6.16)$$

Clearly, for the bonding MOs a value of $\alpha = 0$ localizes the electron completely on the metal, $\alpha = 1$ localizes the electron on the ligands, and $\alpha = \pm 1/\sqrt{2}$ shares the electron equally between the metal and the ligands. On the other hand, for the antibonding MOs, $\alpha = 1$ localizes the electron on the metal, $\alpha = 0$ on the ligands, and again $\alpha = \pm 1/\sqrt{2}$ shares the electron between metal and ligands. In other words, the bonding electrons are localized on the ligands while the antibonding electrons are on the metal to the *same extent* if $\alpha \sim 1$ or *vice versa* if $\alpha \sim 0$. Thus, if both the bonding and antibonding molecular orbitals are filled there will be no net transfer of electron charge between ligands and metal, irrespective of the value of the mixing coefficient α. In this sense we say there is no net covalency contribution to the *bonding* from the orbitals.

We now discuss the origin of the observed covalency in the NiF_6^{4-} cluster in the light of the above argument. The antibonding t_{2g} and e_g MOs contain the eight 'metal 3*d* electrons' giving the $(t^*_{2g})^6(e^*_g)^2$ configuration† while the corresponding bonding MOs are completely filled by the essentially fluoride 2*s* and 2*p* electrons, giving $(t_{2g})^6(e_g)^4$. It is clear that the completely filled bonding–antibonding pair of t_{2g} orbitals does not make a covalency contribution to the bonding. Also, since in an open shell system a distinction is made between up-spin and down-spin electrons, we can further eliminate the covalency contribution from the e_g and e^*_g electrons of the same spin. As discussed by Watson and Freeman, the sole covalency contribution to $10Dq$ comes from the e_g bonding electrons of 'unpaired' or minority spin. We summarize this argument in figure 6.3 which is a schematic representation of the relevant orbitals taken from the complete MO energy level diagram.

We see that the pair of bonding and antibonding t_{2g} MOs gives no net

† We now use the conventional asterisk to denote an antibonding orbital.

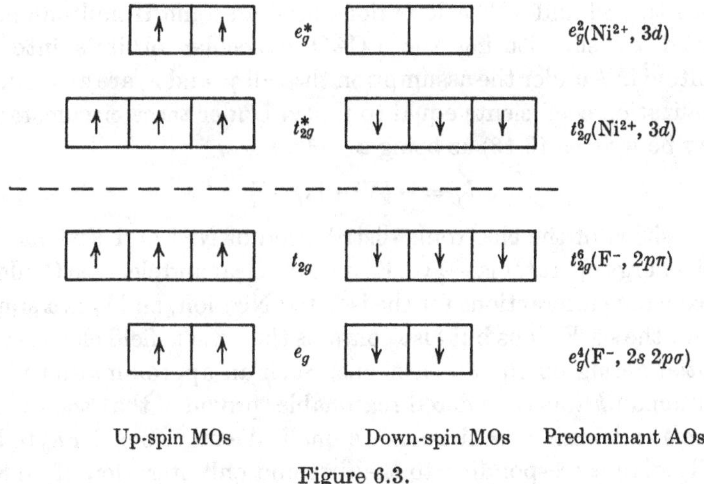

Figure 6.3.

covalency contribution for both up- and down-spin orbitals; likewise for the up-spin pair of e_g bonding and antibonding MOs. The observed covalency effects have their source, therefore, in the 'unpaired' (WF terminology) e_g bonding electrons of minority spin. However, we must emphasize that the paired bonding and anti-bonding electrons still contribute to the crystal-field splitting by virtue of the *overlap and non-orthogonality* of metal and ligand orbitals, as we shall see. Covalency effects may be studied using the bonding electrons which have no antibonding partners or, alternatively, the antibonding holes – the unoccupied (virtual) antibonding orbitals. Sugano and Shulman considered the latter alternative but used a Hartree–Fock approximation that is valid only for occupied orbitals. Watson and Freeman used the bonding orbitals and obtained values for the γ coefficients.

In these one-electron calculations, the crystal-field splitting is given by the orbital energy difference:

$$10Dq = \langle\phi_{ex}|h|\phi_{ex}\rangle - \langle\phi_{gd}|h|\phi_{gd}\rangle, \tag{6.17}$$

where ϕ_{ex} and ϕ_{gd} are wavefunctions appropriate to the excited and ground energy states, respectively, and h is the single electron Hamiltonian for the system. In the present case, ϕ_{ex} is an e_g antibonding MO and ϕ_{gd} is a t_{2g} antibonding orbital. The orbital energies are obtained as solutions to the suitably chosen Hamiltonian, as are the MO wavefunctions in terms of the γ_i and λ_i of (6.10). The first step, therefore, is the determination of a satisfactory form for the Hamiltonian.

Both the SS and WF calculations used an ionic Hamiltonian, h_0, obtained by substituting the LCAO molecular orbitals into the Hamiltonian h under the assumption that all γ_i and λ_i are zero and all normalization coefficients equal to unity. Under these circumstances h_0 may be written (6.18) as being a

$$h_0 = -\tfrac{1}{2}\nabla^2 + V_M + V_L \tag{6.18}$$

superposition of the electronic distribution of Ni^{2+} and F^- ions. The kinetic energy operator is $-\frac{1}{2}\nabla^2$; V_M is the nuclear and electron Coulomb and exchange interactions for the isolated Ni^{2+} ion; and V_L is a similar term for the six F^- ions but also contains the crystal-field electrostatic potential acting on the $3d$ electrons. Such an approximation to the Hamiltonian h was considered reasonable, provided that the overlap integrals and mixing coefficients are small. We may consider h_0 to be a Hamiltonian corresponding to the first and only iteration of an SCF calculation as it is not corrected by including the S_i and γ_i calculated from it.

Corrections to this ionic Hamiltonian for overlap and covalency effects were considered by Watson and Freeman who also commented on the changes in electron density distribution which result thereby. The overlap contribution, being essentially due to metal–ligand non-orthogonality was considered in the limit of zero covalency by setting $\lambda_i = S_i$ and $\gamma_i = 0$. This was seen to shift charge away from the overlap region onto the metal and ligand ions. The often quoted transfer of charge from ligand orbitals to essentially metal orbitals as a result of orbital *overlap* is thus invalid. On the other hand, the covalency contributions were obtained by evaluating the total overlap contribution for non-zero γ_i and $\lambda_i = \gamma_i + S_i$ and subtracting off the previously considered non-orthogonality part. For this covalency effect, members of an occupied pair of bonding and antibonding MOs cancelled one another, leaving only a contribution from the unpaired bonding electrons. These contributions lead to a donation of electron charge from the ligands to the metal and overlap regions. These interpretations of the effects of overlap and covalency illustrate the important point that the covalent mixing in a pair of bonding and antibonding orbitals conserves charge on the metal whereas the mixing in of an unpaired bonding orbital shifts charge from the ligand to the metal. It is this latter effect that contributes to observables associated with covalency.

The value of $10Dq$ calculated by Watson and Freeman is given by

$$10Dq = E[(t_{2g}^*)^5(e_g^*)^3]_{\text{ex}} - E[(t_{2g}^*)^6(e_g^*)^2]_{\text{gd}}, \tag{6.19}$$

that is, the difference between the energies E of excited (ex) and ground (gd) states, these states differing in their highest configurations as shown. Alternatively, we can write $10Dq$ as the difference in energies ϵ of the orbitals involved in the crystal-field transition:

$$10Dq = \epsilon[e_g^*(x^2-y^2)\downarrow] - \epsilon[t_{2g}^*(xy)\downarrow] \quad (6.20)$$

where the down-spin (minority) electrons are involved for reasons given earlier and the crystal-field transition is assumed to occur between the x^2-y^2 and xy components of the e_g and t_{2g} MOs. The e_g and t_{2g} MOs are those derived from the two different states but if we assume constant orbital behaviour in these states then the orbitals may be described using the ground state Hamiltonian. In the case of the NiF_6^{4-} complex, this procedure retains the RHF method for closed shells since the ground state $t_{2g}\downarrow$ orbital is filled and the ground state $e_g\downarrow$ orbital is empty. Within this one-electron formalism, the total energy of the ground state is related to the sum of the orbital energies. The value of $10Dq$ is given by the differences in the orbital energies for all such energies that are changed as a result of the crystal-field transition.

In the SS calculation, the bonding electrons are unchanged by this transition; in fact their presence is admitted only insofar as providing orthogonal partners for the antibonding electrons. The WF calculation, because of its argument concerning the active role played by these same bonding electrons, considered changes in these bonding electrons as well as those directly involved in the transition. It transpires that these changes determine the precise nature of the contributions to $10Dq$ arising from the various orbital energies. Figure 6.4 shows the molecular-orbital electronic configurations before and after the transition. The sets of MOs are labelled by their symmetry designations and by their predominant atomic character. The spin subdivision is omitted here and the two spin orbitals of each symmetry type are implicit in the box representation. We note that the transition changes the $\zeta\downarrow$ electron from being 'paired' with the $\zeta^*\downarrow$ electron in the ground state to an 'unpaired' situation in the excited state. Similarly, the 'unpaired' $v_\sigma\downarrow$ and $v_s\downarrow$ electrons in the ground state become paired with the $v^*\downarrow$ in the excited state. Thus, the roles of the bonding–antibonding pairs and of unpaired bonding electrons may change from the ground to the excited state and so, therefore, do their contributions to $10Dq$. The electrons which do change roles on transition are tabulated in table 6.1 where is shown also, the nature of their contribution to $10Dq$.

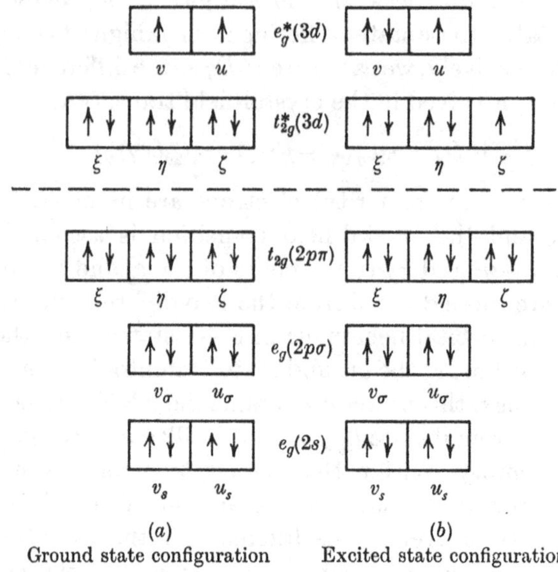

Figure 6.4.

The table summarizes how the covalent contributions to $10Dq$ arise from the bonding electrons of minority spin and how the antibonding electrons make only overlap (and diagonal) contributions. Watson and Freeman therefore considered the one-electron orbital energies which give non-zero contributions to $10Dq$ and hence obtained the energies of the ground and excited states as sums of the orbital energies for those orbitals shown in table 6.1. In this way they obtained a complicated equation for $10Dq$ which was used not only to compute the value

TABLE 6.1. *Contributions to* $10Dq$ *in one-electron calculations for octahedral* d^8 *systems*

Ground state contribution	MO	Excited state contribution
Overlap	$\zeta^* \downarrow$	None
Overlap	$\zeta \downarrow$	Covalency
Covalency	$v_\sigma \downarrow$	Overlap
Covalency	$v_s \downarrow$	Overlap
Covalency	$u_\sigma \downarrow$	Covalency
Covalency	$u_s \downarrow$	Covalency
None	$v^* \downarrow$	Overlap

TABLE 6.2. *Breakdowns of contributions to $10Dq$ for NiF_6^{4-}, calculated by one-electron and ab initio calculations*

Origin		Contribution to $10Dq$ (cm^{-1})				
		one-electron calculations			*ab initio*	
		SS[97]		WF[c]	Richardson *et al.*[d]	
Diagonal	V_L^0	+1390			+1532	
	V_L^K	−2080	−2670	−3570	−1129	−2823
	V_L^E	−2880			−3226	
	Renormalization	+900				
Non-orthogonality (overlap)		+3730		+5725	+7501	
Covalency		+5290		+660	+3307	
Total calculated		6350		2815	7985	
Experimental		7250				

of $10Dq$ but also to analyse the separate contributions from the overlap, covalency and diagonal crystal-field terms.

The total value of $10Dq$ for NiF_6^{4-} was calculated to be *ca.* 2800 cm^{-1}. The separate contributions are shown in table 6.2 where we also list the corresponding results obtained by Sugano and Shulman and, for later comparison, by Richardson *et al.* The diagonal contributions arise from the matrix elements of the ionic Hamiltonian under the condition, $\gamma_i = 0$, $S_i = 0$, when we have:

$$\langle\psi_i|h_0|\psi_i\rangle = \langle\phi_i|h_0|\phi_i\rangle = \langle\phi_i|-\tfrac{1}{2}\nabla^2+V_M+V_L|\phi_i\rangle$$
$$= \langle\phi_i|-\tfrac{1}{2}\nabla^2+V_M|\phi_i\rangle+\langle\phi_i|V_L|\phi_i\rangle, \qquad (6.21)$$

where ϕ_i are the metal $3d$ orbitals. The first term on the right-hand side of (6.21) is the orbital energy of the $3d$ electrons and the second term is the increase in this energy in the ligand field. The electrostatic value of $10Dq$ is then given by $(10Dq)_0$, where

$$(10Dq)_0 = \langle\phi_{e_g}|V_L|\phi_{e_g}\rangle - \langle\phi_{t_{2g}}|V_L|\phi_{t_{2g}}\rangle. \qquad (6.22)$$

The explicit form of V_L can be written, following Sugano and Shulman, as

$$V_L = V_L^0 + V_L^K + V_L^E, \qquad (6.23)$$

where V_L^0 is the well-known point-charge or point-dipole potential in simple crystal-field theory, V_L^K is Kleiner's correction resulting from the interaction of the metal electrons with the ligand nucleus and with the ligand electron clouds expressed by the Coulomb interaction, and

V_L^E is the potential arising from the exchange interactions between metal and ligand orbitals first introduced by Tanabe and Sugano. The matrix elements expressed as the second term in (6.21) thus constitute the purely ionic model (only a part of which arises from the point-charge model) and the contribution from this source was found to be $-3570\,cm^{-1}$ by Watson and Freeman, and $-2670\,cm^{-1}$ by Sugano and Shulman. Clearly, *the ionic model cannot by itself account for the crystal-field splittings either in terms of magnitude or sign*. As pointed out by Sugano and Shulman, the ionic model taken to its logical limits predicts a negative Dq value; further, this conclusion is independent of the form of the 3d wavefunctions. The more expanded the orbital, the greater is the point-charge contribution but so also are the negative contributions from Kleiner's penetration correction and from the exchange terms. From this we see that the fortuitious agreements obtained by Van Vleck and by Polder result from the use of over-expanded 3d orbitals together with neglect of V_L^K and V_L^E.

TABLE 6.3. *10Dq values for some fluoride complexes, calculated by the WF method*

Complex ion	10Dq (calculated)	10Dq (observed)	Reference
MnF_6^{4-}	$3240\,cm^{-1}$	*ca.* $8300\,cm^{-1}$	Gondaira[48]
FeF_6^{3-}	7300	14000	Offenhartz[76]
CrF_6^{3-}	6800	*ca.* 15200	Offenhartz[76]

The various contributions from the overlap or non-orthogonality and covalency terms correct the sign of Dq and so re-establish agreement with experiment regarding the manner of d orbital splittings if not their magnitudes. The WF and SS calculations differ greatly in the relative magnitudes of these two effects. Watson and Freeman find that overlap dominates covalency while Sugano and Shulman find the opposite result. The WF model has been used by other workers with no better results. The calculated values obtained for 10Dq shown in table 6.3 for some fluoride complexes are generally too low by a factor of two.

Finally, Watson and Freeman were pessimistic as to whether the LCAO–MO method was capable of yielding a better result even if the exact Hamiltonian were used. Also, the constraints imposed by this method could not allow many vital questions to be answered. For example, does the one-electron orbital picture lead to results consistent with excitations between energy states? To what extent are the atomic orbitals changed in going from the free-ion to the complex environment? Before passing onto a consideration of these problems within the SCF–MO framework, we mention briefly an alternative to the molecular–orbital formalism.

THE HEITLER–LONDON APPROACH

The weak covalency in metal complexes has been considered from the view of the Heitler–London or valence-bond method by Hubbard, Rimmer and Hopgood.[54] The covalency is introduced by mixing configurations from 'excited states' corresponding to electron transfer from the ligands to the central metal. As a crude approximation we can write the total many-electron wavefunction for, say, NiF_6^{4-} as

$$\Psi = \Psi(Ni^{2+}F_6^-) + a\Psi(Ni^+FF_5^-), \qquad (6.24)$$

where the second term represents a configuration in which a fluorine electron has been transferred to a metal orbital and a is the mixing coefficient. The important advantage of this configuration interaction (CI) model in comparison to the LCAO–MO scheme is that it allows the relaxation of some of the constraints on orbital behaviour. Thus the $3d$ orbitals expand on going from Ni^{2+} to Ni^+, and conversely the fluorine orbitals contract as F^- changes to F. These effects allow for a correlation of the d electrons with the added electrons and also for changes in mutual polarization of the metal and ligand ions in the complex. The approach allows for the physical changes that undoubtedly occur in the electron behaviour on complexation in a way not amenable to the previously considered MO approach. That is, the zeroth-order of the valence-bond method corresponds more nearly with chemical intuition than does the zeroth-order MO model. Actually better agreement between the calculated 5400 cm^{-1} and observed values of $10Dq$ for NiF_6^{4-} has been obtained by the valence-bond calculation.

However, despite this more successful calculation, the advantage of the CI approach over the MO one depends more upon the approximations adopted in performing calculations using the latter than on any intrinsic qualities. The various correlation effects could be incorporated into the MO formalism if a sufficiently large basis set of functions were used. As the most recent work has concentrated upon the MO rather than the CI approach, we return to this topic in the next section.

MANY-ELECTRON MO CALCULATIONS

The most recent calculations of the crystal-field energy levels have employed the Hartree–Fock SCF molecular-orbital theory and many, if not all, of the electrons in the complex ion have been considered. The desirability of these lengthy calculations was indicated by the problems discussed earlier and they have become technically feasible owing to modern developments in computer technology. It should be stressed, however, that many approximations are involved in such calculations and some care must be exercised in any physical interpretations put on the results: further, calculations based on different assumptions are often amenable only to qualitative comparison. The various calculations reviewed in this section are summarized in table 6.4 showing computed spectral transition energies. In almost all the cases considered the subject is the NiF_6^{4-} ion.

The first calculation in this category is that by Ellis, Freeman and Ros[22] who used a theoretical model based on the unrestricted Hartree–Fock formalism. All matrix elements were computed with the aid of a one-centre basis set of molecular wavefunctions composed of variable proportions of Slater functions for the Ni^{2+} and F^- ions. As a consequence of allowing different spin behaviour, appreciable spin-splitting of the $3d$ orbitals occurs which was found to be greater for the 'open-shell' metal e_g electrons than for the 'closed-shell' t_{2g} electrons of the $^3A_{2g}$, $t_{2g}^6e_g^2$ energy state. This behaviour makes comparison with traditional ligand-field theory difficult. This difficulty is augmented by the orbital rearrangements that are considered to occur on excitation between the separately computed $^3A_{2g}$ and $^3T_{2g}$ states so that the excitation energy cannot be treated simply as the one-electron promotion energy $\epsilon(t_{2g}) \rightarrow \epsilon(e_g)$. The transition energy was found to be in reasonable agreement with experiment although the authors' comment that this value represents a difference in the sixth significant figure of the total energies should be borne in mind.

A restricted Hartree–Fock calculation for NiF_6^{4-} by Gladney and Veillard[47] has used the approximation of a minimum basis set of GTOs for each of the Ni^{2+} and F^- orbitals. The total energies of the $^3A_{2g}$, $^3T_{2g}$, $^3T_{1g}$ and $^1T_{1g}$ states were calculated and the results analysed in terms of the parameters $10Dq$, B and C. The computed value of $10Dq$ was some 30 % below the observed value and the interelectron repulsion parameters were found to be substantially the same as those calculated for the free Ni^{2+} ion. The interesting point about the latter figure is their consistency with almost free-ion electron densities: thus, for the $^3A_{2g}$ state the net charges were calculated as $Ni^{1.9594+}$ and $F^{0.9932-}$ so that only a small fraction of charge is transferred from the ligands to the metal.

RHF and UHF procedures were compared by Moskowitz *et al.*[69] in a calculation of the energies of the $^3A_{2g}$ $(t_{2g}^6e_g^2)$ state and the excited $^3T_{1g}$ and $^3T_{2g}$ $(t_{2g}^5e_g^3)$ states, again for the NiF_6^{4-} complex. A large basis set of GTO functions was used for all 86 electrons of the cluster, together with empty 4*s* and 4*p* orbitals of the Ni^{2+} ion: this selection was hoped to alleviate problems arising from the choice of basis functions. As with the previous UHF calculation, a pronounced spin-splitting of the t_{2g} and e_g molecular orbitals occurred so that a simple comparison of the two procedures was not possible. Despite this complication, the RHF and UHF calculations were equally good in describing the electronic properties of the complex. However, it was shown that a rearrangement of the excited state MOs occurring on excitation precludes any realistic calculation of $10Dq$ using the orbital energies determined solely by the ground state calculation. Such a calculation actually yielded a value of 33 250 cm^{-1} for $10Dq$!

A procedure described by Richardson and his co-workers[60, 89, d] has had some success in calculating the electronic energy levels and spectra of various metal hexafluoride complexes. The theoretical model uses a 'frozen core' approximation in which the inner orbitals (Ni^{2+} 1*s*, 2*s*, 2*p* and F^- 1*s*) are regarded as free-ion functions having a negligible effect on the transition energies. The various state energies arising from the $t_{2g}^6e_g^2$ and $t_{2g}^5e_g^3$ configurations were calculated using RHF and UHF procedures in which all one- and two-centre integrals were computed but three- and four-centre integrals were approximated or ignored. The parameter $10Dq$ is correlated with the difference in the energies of two independently calculated many-electron states, the value shown in table 6.4 being in excellent agreement with experiment. The monotony of calculations limited to fluoride ion complexes was broken by an *ab initio* calculation of the electronic structure and spectrum of $Co(NH_3)_6^{3+}$ performed by Kalman and Richardson.[60] The 'frozen core' approximation was used and the results for the ammonia complex were compared with those for the CoF_6^{3-} complex, similarly calculated, to emphasize the different splitting effects of the two ligands. The values of $10Dq$ shown in table 6.4 were obtained as differences between the ground and first excited states for the high-spin $t_{2g}^4e_g^2$ ground configuration and two of the excited states for the low-spin t_{2g}^6 ground configuration in $Co(NH_3)_6^{3+}$. The computed electronic population in the ground energy state suggested a transfer of 1.1

TABLE 6.4. *Many-electron SCF–MO calculations of* 10*Dq*†

Dq†(cm^{-1})		Complex ion	Reference
Calculated	Observed		
10800	7250	NiF_6^{4-}	Ellis *et al.* [22]
4670	7250	NiF_6^{4-}	Gladney and Veillard[47]
17150	17060	TiF_6^{3-}	Richardson *et al.*[89]
15879	16100	CrF_6^{3-}	Richardson *et al.*[89]
8090	7660	FeF_6^{4-}	Richardson *et al.*[89]
7126	7250	NiF_6^{4-}	Richardson *et al.*[89]
~ 5000	~ 8000	MnF_6^{4-}	Matsuoko[68]
6089	7250	NiF_6^{4-}	Moskowitz *et al.*[69]
7210	7250	NiF_6^{4-}	Richardson *et al.*[d]
12124	12986	CoF_6^{3-}	Kalman and Richardson[60]
28232	~ 23000	$Co(NH_3)_6^{3+}$	Kalman and Richardson[60]

† The calculated values are correlated with the term energy differences computed for the first spin-allowed ligand-field transitions, except for $Co(NH_3)_6^{3+}$ for which ${}^5T_{2g} \rightarrow {}^5E_g$ was used.

electrons from the NH_3 ligands to the cobalt atom and 0.25 electrons from the fluorines to cobalt. Neither of these electron transfers are sufficient to effectively neutralize the cobalt atom and in this sense they cast some doubt on the applicability of the electroneutrality principle.

While these many electron approaches currently constitute the best available means of calculating the energy states of transition-metal complexes, they do not lend themselves readily to an interpretation of 10*Dq* as a difference in orbital energies consistent with the spirit of ligand-field theory. In particular, they offer no indication of the various factors contributing to this parameter. To answer these difficulties Richardson has analysed his results for the NiF_6^{4-}, CoF_6^{3-} and $Co(NH_3)_6^{3+}$ complexes to obtain single-electron orbital descriptions and to examine the plausibility of such descriptions. The relative energies of the ${}^3A_{2g}$ and ${}^3T_{2g}$ state in NiF_6^{4-}, for example, were determined for each successive refinement to the point-charge crystal-field model so determining the various contributions to 10*Dq*. The breakdown is shown in table 6.2. It is interesting to observe how the contributions from orthonormalization and covalency are much closer to the (incorrectly determined) values obtained by Shulman and Sugano than those of Watson and Freeman.

ORBITAL ENERGIES AND DEFINITIONS OF 10Dq

No doubt it would be desirable to calculate 10Dq values for an extensive range of transition-metal and lanthanide complexes using all-electron (or perhaps, 'frozen core') *ab initio* SCF techniques. If so it would be equally rewarding to calculate spectral splitting patterns for distorted systems too and hence compute values for Cp and all the other parameters discussed in this book. This is clearly a formidable and expensive task and one unlikely to be undertaken in the near future. Until that time, various semi-empirical methods will continue to be employed and some of these form the subject matter of the next two chapters. However, before leaving our review of the 'history of 10Dq' it is worth clarifying one important aspect of the one-electron approximation we have already discussed.

In chapter 1 and in many elementary texts, essentially three notions of the definition of 10Dq arise. Octahedral symmetry simply defines 10Dq in an ion with a d^1 configuration as the difference in energy between the ^2E_g and $^2T_{2g}$ terms which, in this case, is exactly the same as the energy difference between putting the one electron either in an e_g orbital or in a t_{2g} orbital. As a definition we can equate this energy difference with the energy difference between the e_g and t_{2g} *orbitals*. The situation is a little different for, say, d^2 ions, however. Recalling our discussion in chapter 1, we may define 8Dq as the energy separation between the $^3T_{2g}$ and $^3T_{1g}(F)$ terms, for example, or in order to generate comparability through the d^n configurations, we may define 10Dq as the energy difference between t_{2g}^2 and $t_{2g}^1e_g^1$ configurations. The 'three definitions' thus refer to term energy differences, configuration energy differences and orbital energy differences. While the definitions are one and the same, problems can arise when computing orbitals as Dahl and Ballhausen[b,e] and also Offenhartz[a,77] have discussed. As we wish to focus attention on orbitals and configurations we first dispose of the term energy separation. If we define (by symmetry) the $^3T_{2g}-{}^3T_{1g}$, or an equivalent, term separation by 8Dq, then this quantity will be given properly by the sort of *ab initio* calculations described in the previous section. Energies are calculated for the complex ion in its ground and various excited states and energy separations corresponding to observable spectral bands are obtained by subtraction. State, as opposed to orbital, energies are calculated here.

However, let us re-examine calculations of orbital energies. The conventional Hartree–Fock formalism gives the orbital energies ϵ_i

as solutions of (6.1) according to the expression:

$$\epsilon_i = H_i + \sum_j (2J_{ij} - K_{ij}), \tag{6.25}$$

where H_i are the matrix elements of $\mathscr{H}$ and are the one-electron terms; J_{ij} and K_{ij} are the two-electron Coulomb and exchange integrals, the sum being over all doubly occupied orbitals j. In an open-shell configuration, however, for a free-ion the orbital energies are subject to the occupancy of these orbitals so that different energies may be associated with the 3d shell in a free transition metal ion. We illustrate this with the $t_{2g}^6 e_g^2$ electronic configuration of the Ni^{2+} ion. Table 6.5 lists the various Coulomb and exchange integrals within the real d orbital set as basis in terms of both Condon and Shortley and of Racah's notations as developed in chapter 3. Assuming that all d orbitals have the same value of H_i, we may compute the interaction of, say, the d_{z^2} orbital containing an up-spin electron with all other functions relevant to the $t_{2g}^6 e_g^2$ configuration. These are shown in table 6.6. Note that the Coulomb and exchange interactions cancel for the electron interacting with itself when $J_{ii} = K_{ii}$; also, exchange interactions obviously occur only between electrons of the same spin (see chapter 3). Similar procedures for up- and down-spin t_{2g} orbitals and empty e_g orbitals lead to the following:

$$\left.\begin{aligned} \epsilon(t_{2g}\uparrow) &= H_i + 7A - 14B + 5C, \\ \epsilon(t_{2g}\downarrow) &= H_i + 7A - 10B + 7C, \\ \epsilon(e_g\uparrow)^f &= H_i + 7A - 14B + 3C, \\ \epsilon(e_g\downarrow)^e &= H_i + 8A - 6B + 7C, \end{aligned}\right\} \tag{6.26}$$

where the superscripts e and f refer to empty or filled e_g orbitals. This exercise shows how the orbital energies depend on occupancy: further these energies refer only to the $t_{2g}^6 e_g^2$ configuration. The energies for these same orbitals would be different if calculated for the states of the $t_{2g}^5 e_g^3$ configuration. The primary reason why the energy of the empty e_g orbital differs from those of the filled orbitals arises from the self-cancellation of J and K for an electron interacting with itself. Thus, an electron in a filled orbital interacts with or 'sees' only seven electrons (in the present case) whereas the empty orbital sees all eight electrons. Notice how orbital energy differences occur in the free-ion although an externally perturbing crystal-field has not yet been introduced.

It is worth considering this notion further. The preceding discussion was concerned with a free-ion configuration and hence with spherical symmetry. Clearly the choice of orbitals defined by the cubic har-

TABLE 6.5. *Two-electron integrals for real d orbitals*

d orbitals		Coulomb integrals $\langle ab\|.\|ab\rangle$		Exchange integrals $\langle ab\|.\|ba\rangle$	
z^2	z^2	$F_0+4F_2+36F_4$	$A+4B+3C$	$F_0+4F_2+36F_4$	$A+4B+3C$
z^2	x^2-y^2	$F_0-4F_2+6F_4$	$A-4B+C$	$4F_2+15F_4$	$4B+C$
z^2	xy	$F_0-4F_2+6F_4$	$A-4B+C$	$4F_2+15F_4$	$4B+C$
z^2	xz	$F_0+2F_2-24F_4$	$A+2B+C$	F_2+30F_4	$B+C$
z^2	yz	$F_0+2F_2-24F_4$	$A+2B+C$	F_2+30F_4	$B+C$
x^2-y^2	x^2-y^2	$F_0+4F_2+36F_4$	$A+4B+3C$	$F_0+4F_2+36F_4$	$A+4B+3C$
x^2-y^2	xy	$F_0+4F_2-34F_4$	$A+4B+C$	$35F_4$	C
x^2-y^2	xz	$F_0-2F_2-4F_4$	$A-2B+C$	$3F_2+20F_4$	$3B+C$
x^2-y^2	yz	$F_0-2F_2-4F_4$	$A-2B+C$	$3F_2+20F_4$	$3B+C$
xy	xy	$F_0+4F_2+36F_4$	$A+4B+3C$	$F_0+4F_2+36F_4$	$A+4B+3C$
xy	xz	$F_0-2F_2-4F_4$	$A-2B+C$	$3F_2+20F_4$	$3B+C$
xy	yz	$F_0-2F_2-4F_4$	$A-2B+C$	$3F_2+20F_4$	$3B+C$
xz	xz	$F_0+4F_2+36F_4$	$A+4B+3C$	$F_0+4F_2+36F_4$	$A+4B+3C$
xz	yz	$F_0-2F_2-4F_4$	$A-2B+C$	$3F_2+20F_4$	$3B+C$
yz	yz	$F_0+4F_2+36F_4$	$A+4B+3C$	$F_0+4F_2+36F_4$	$A+4B+3C$

TABLE 6.6

	$J-K$	Σ
$\langle d_{z^2}\uparrow\|.\|d_{yz}\uparrow\rangle$	$(A+2B+C)-(B+C)$	$A+B$
$\|d_{yz}\downarrow\rangle$	$(A+2B+C)\quad 0$	$A+2B+C$
$\|d_{xz}\uparrow\rangle$	$(A+2B+C)-(B+C)$	$A+B$
$\|d_{xz}\downarrow\rangle$	$(A+2B+C)\quad 0$	$A+2B+C$
$\|d_{xy}\uparrow\rangle$	$(A-4B+C)-(4B+C)$	$A-8B$
$\|d_{xy}\downarrow\rangle$	$(A-4B+C)\quad 0$	$A-4B+C$
$\|d_{z^2}\uparrow\rangle$	$(A+4B+3C)-(A+4B+3C)$	0
$\|d_{x^2-y^2}\uparrow\rangle$	$(A-4B+C)-(4B+C)$	$A-8B$

$$\sum_{\text{spin pairs}} (2J-K) = 7A-14B+3C$$

monics t_{2g} and e_g was an arbitrary matter: spherical harmonics, defining $d_{\pm 2}$, $d_{\pm 1}$, d_0, would have been a more obvious (though no more correct) choice. The point is this: orbitals are no more uniquely defined in a many-electron system than are orbital energies. If electron–electron interactions (the two-electron terms J and K) are ignored, then the d orbitals *may* be uniquely defined (depending on the symmetry, of course) but they will be degenerate. Otherwise orbital, as opposed to state or configuration, energies are not observables and have calculated energies which depend on the (arbitrary) way in which the functions are defined.

Under the influence of a crystal field, the orbital energy becomes:

$$\epsilon_i = H_i + \sum_j (2J_{ij} - K_{ij}) + V_L, \tag{6.27}$$

where V_L is the energy of the ith orbital arising from the crystal-field potential; i.e. $V_L = \langle \phi_i | V_L | \phi_i \rangle$. If we define $10Dq$ as the difference in the energies of the t_{2g} and e_g orbitals in the crystal field, we have

$$\begin{aligned} \epsilon(e_g) - \epsilon(t_{2g}) &= \text{`}10Dq\text{'} \\ &= V(e_g) - V(t_{2g}) + [\Sigma(2J-K)_e - \Sigma(2J-K)_{t_{2g}}]. \end{aligned} \tag{6.28}$$

Hence, if the orbital energies are defined in the usual way, $10Dq$ is not simply given by the energy differences in the crystal field, but includes differences in two-electron terms. Now, if we have in mind a weak-field–strong-field correlation diagram, the foregoing is obvious and, in a sense, trivial. At the extreme of infinitely strong fields, we ignore interelectron repulsion terms and, as discussed in the previous paragraph, orbital energies and configuration energies become the same. The point about (6.28) is that we can still choose to refer to 'orbitals' even in the presence of interelectron repulsion. While they are not observables, they can feature in a molecular-orbital calculation and have energies associated with them. But orbital energies and configuration energies in the circumstances are *not* equivalent and this is what (6.28) expresses.

An MO estimate of the $A_{2g} - T_{1g}$ energy separation in octahedral nickel(II) ions involves an entirely separate calculation from that giving the $A_{2g} - T_{2g}$ separation: this is in contrast to crystal-field theory which essentially relates these quantities by symmetry arguments. An estimate of $10Dq$ as an orbital energy difference from the first MO calculation will generally be quite different from one deduced from the second calculation. Clearly a definition of $10Dq$ which depends on the way it is calculated is meaningless. The best one can do is to *correlate* $10Dq$ from crystal-field and MO calculations. Thus the ratio of $A_{2g} - T_{2g} : A_{2g} - T_{1g}$ splittings from MO calculations are indeed near 10:18 as predicted by crystal-field theory and in this sense we could then identify (but not equate by definition) the $A_{2g} - T_{2g}$ separation so calculated, with $10Dq$. But, of course, if one goes to the trouble of making an MO calculation, there is no real point in calling this energy separation $10Dq$ as all one is really attempting is to calculate a spectral transition energy.

CONCLUSION

In concluding this chapter we note that several of the traditional concepts of the orbital picture of ligand-field theory require changing. Chief among these is the role of the unpaired bonding electrons rather than the antibonding ones in contributing to the covalency in these molecules. Clearly, the bonding in NiF_6^{4-} is associated with the same types of molecular orbitals as those in SF_6. The peculiarity of transition metal complexes is that they also have electrons accommodated in antibonding orbitals and it is perhaps unfortunate that many of the measurable electronic properties are associated with these electrons. However, while many problems are still unresolved in computing *ab initio* values of $10Dq$, it is clear that the major contribution to this quantity comes from the covalent mixing of the metal and ligand orbitals.

GENERAL REFERENCES

(*a*) P. O'D. Offenhartz, *Atomic and Molecular Orbital Theory*, McGraw-Hill, New York, 1970.

(*b*) J. P. Dahl and C. J. Ballhausen, Molecular orbital theories of inorganic complexes, *Advances in Quantum Chemistry*, 1968, **4**, 170.

(*c*) R. E. Watson and A. J. Freeman, Covalency in crystal field theory: $KNiF_3$, *Phys. Rev.* 1964, **134**, A1526.

(*d*) T. F. Soules, J. W. Richardson and D. M. Vaught, Electronic structure and spectrum of the NiF_6^{4-} cluster: results of calculations based on self-consistent-field models, *Phys. Rev. B.* 1971, **3**, 2186.

(*e*) C. J. Ballhausen, Crystal and ligand field theory, *Int. J. Quantum Chem.* 1971, **5**, 373.

7
A CRYSTAL-FIELD APPROACH TO RADIAL PARAMETERS

We devoted some space in our opening chapter to separating symmetry-based aspects of ligand-field theory from other 'quantitative' ones. While some chemically interesting information may be obtained from studies of only the symmetry properties of ligand-field models, our interest generally lies in properties which depend upon the magnitudes of the radial parameters. Ultimately it is these quantities which permit the comparisons between systems, with respect to metal, ligand and geometry, which characterize chemistry. The Spectrochemical Series with its property of factorizability into functions of metal and ligand is a typical 'chemical' concept with limited appeal for those more concerned with 'fundamentals'. The calculation of absolute *Dq* values has enjoyed a long history, as we saw in the previous chapter. Whatever else, that discussion showed that the representation of ligands by point-charges or point-dipoles without explicit recognition of chemical bonding in calculations is wholly inadequate for the determination of theoretical values for *Dq*. But complete calculations of *Dq* and other radial parameters are difficult and generally too lengthy to be seriously contemplated for all combinations of metal and ligand. The same comment applies throughout chemistry, of course. Until the advent (if ever) of enough theoretical understanding to construct a computer program, and of a computer fast enough to carry it out, that will calculate all properties and reactions of all molecules, chemistry must rely on empirical or, at least, semi-empirical models. A common characteristic of such models is that they may relate observations in one sphere of chemistry without having relevance in another. From a chemical point of view such models must be valued for their utility and their ability to suggest new experiments rather than for their relation to an all-pervading 'truth'. As far as ligand-field radial parameters are concerned, we shall discuss two different semi-empirical approaches representing essentially electrostatic or covalent bonding as zeroth-order approximations. In the next chapter we study a semi-empirical molecular-orbital method called the 'angular overlap' method. In this chapter we see how far the apparently disreputable point-charge (or point-dipole) model may be stretched.

Recalling the small contribution made by electrostatic repulsion terms in a modern *ab initio* calculation of $10Dq$, and indeed that inclusion of penetration effects (Kleiner's correction) and exchange terms inevitably cause electrostatic contributions to $10Dq$ to be negative, it might seem pointless to persue the point-charge approach at all. There are really two reasons for doing so. First we have the observation, noted in the previous chapter, that the original calculations of Dq by Polder[85] with point-charges or by Van Vleck[113] with point-dipoles, both using metal radial functions described in terms of the 'over-expanded' Slater-like orbitals, *coincidentally* gave answers in fair agreement with experiment. Similar calculations by Ballhausen and Ancmon[4] of second-order radial parameters were able to reproduce some spectral observations from distorted octahedral copper(II) molecules quite well. Thus in a few cases at least, a model which must be described as 'unrealistic' has found some utility. The second reason for studying the point-charge approach further is its great simplicity. In this chapter we examine the model in some detail to see if it possesses a general utility throughout the d (and f) block elements.

There are several points which the model should clarify. These include the various aspects of the Spectrochemical Series including the increasing size of Dq with increasing oxidation state or with principal quantum number in the d-block. Also of interest is the way Dq as defined [in (2.69)] is roughly independent of coordination number while the ratio Cp/Dq is not. The factorizability of the Spectrochemical Series into functions of only metal and only ligand requires comment. And, of course, predictions about the magnitudes of Cp are required, as also a rationalization of any inverse relation between trends in Dq and in Cp/Dq, as discussed in chapters 4 and 5. Our approach is exploratory. We are concerned with trends in parameter values and insight into the nature of these parameters rather than in accurate calculations which, by the nature of the approximations and assumptions involved, would be specious.

THE RADIAL INTEGRALS G^l

We must begin this study by correcting something we wrote in chapter 2 and only implicitly put right in chapter 3. This concerns the formula for the expansion of the inverse distance between a metal electron and a point-charge ligand, in the case of crystal-field theory. The correct expression is:

$$\sum_i \frac{1}{r_{12}} = \sum_i \sum_{l=0}^{\infty} \sum_{m=-l}^{l} \frac{4\pi}{2l+1} \cdot \frac{r_<^l}{r_>^{l+1}} \cdot Y_l^m(\theta_2, \phi_2) \,.\, Y_l^{m*}(\theta_1, \phi_1), \qquad (7.1)$$

where we sum over i ligands for a representative point coordinate of the metal electron (2). For all cases when the representative point (2) lies inside the 'coordination shell', i.e. $r < a$, where a is the bond length, we can replace $r_<$ by r and $r_>$ by a. Hence:

$$\sum_i \frac{1}{r_{12}}(r < a) = \sum_i \sum_{l=0}^{\infty} \sum_{m=-l}^{l} \frac{4\pi}{2l+1} \cdot \frac{r^l}{a^{l+1}} \cdot Y_l^m(\theta_2, \phi_2) \,.\, Y_l^{m*}(\theta_1, \phi_1), \tag{7.2}$$

which is the equation used in chapter 2. However, we should additionally take those cases where the representative point (2) is outside the coordination shell, in which case $r_< = a$ and $r_> = r$, so that we get:

$$\sum_i \frac{1}{r_{12}}(r > a) = \sum_i \sum_{l=0}^{\infty} \sum_{m=-l}^{l} \frac{4\pi}{2l+1} \cdot \frac{a^l}{r^{l+1}} \cdot Y_l^m(\theta_2, \phi_2) \,.\, Y_l^{m*}(\theta_1, \phi_1). \tag{7.3}$$

Of course (7.1) is the sum of (7.2) and (7.3). This point was not made in chapter 2 simply to avoid complicating our main theme there. Updating the rest of the material in that and subsequent chapters is a straightforward procedure. There are no changes required in anything except the definitions of the radial integrals Dq and Cp and in anything subsequently expressed in those terms. Thus, whenever we see

$$\int_0^{\infty} R(d) \,.\, \frac{r^l}{a^{l+1}} \,.\, R(d) \,.\, r^2 \,.\, dr, \tag{7.4}$$

we should read

$$\int_0^{a} R(d) \,.\, \frac{r^l}{a^{l+1}} \,.\, R(d) r^2 dr + \int_a^{\infty} R(d) \,.\, \frac{a^l}{r^{l+1}} \,.\, R(d) r^2 dr. \tag{7.5}$$

No argument made so far is affected by this change: the total quantity in (7.5) may be regarded simply as 'the radial integral'. We may now formally re-define the radial parameters Dq and Cp, and introduce the integrals G^l:

$$G^l = \int_0^{a} R^2(d) r^2 \,.\, \frac{r^l}{a^{l+1}} \,.\, dr + \int_a^{\infty} R^2(d) r^2 \,.\, \frac{a^l}{r^{l+1}} \,.\, dr, \tag{7.6}$$

$$Dq = \tfrac{1}{6} z e^2 G^4, \quad Cp = \tfrac{2}{7} z e^2 G^2, \tag{7.7}$$

where ze is the effective charge on the ligand and $R(d)$ is the radial part of the metal d wavefunction. These expressions characterize the point-charge representation of ligands.

PICTORIAL REPRESENTATION OF THE G^l INTEGRALS

Although only the total integrals G^l represent observables, it is instructive to show their breakdown into constituent functions diagrammatically. The integrals are the products of the square of the metal wavefunction and a term derived from the potential set up by the point ligands. Taking the radial part of the wavefunction as giving electron density normalized to unity, we have

$$\int_0^\infty R(d).r.dr = 1 \quad \text{and} \quad \int_0^\infty R^2(d).r^2.dr = 1, \tag{7.8}$$

and thus choose to associate the r^2 in (7.5) with the metal rather than with the ligand: this is shown in (7.6). This is an arbitrary matter which makes no real difference to the subsequent discussion, though we shall refer to the matter again shortly. Meanwhile, we can draw a schematic diagram showing the functions under the integral sign in (7.6) as in figure 7.1. Consider the potential term r^l/a^{l+1} when $r < a$ or a^l/r^{l+1} when $r > a$. For any value of l, this term has the value $1/a$ when $r = a$. The curves in figure 7.1, representing second- and fourth-order potentials are simply second- and fourth-power functions of r, scaled by different constants (a^3 and a^5, respectively). Their form ensures that higher-order potentials will be smaller than lower-order ones, except at $r = a$ when they are equal. Physically this means that the effect of the point-charge falls off more rapidly with distance from the point-charge in fourth-order (say) than in second-order. All this is represented by the potential cusps in figure 7.1.

There is a small, but important, point here about the interpretation of the cusps in figure 7.1. Physically, we place point-charges distant a from the metal and these set up a potential which in turn has an effect on the metal electrons. The electric potential set up by the ligand falls off as the square of distance from the ligand, of course. Further, the potential at the position of the unit charge is ze and at a distance x from the charge takes a value ze/x independent of the metal–ligand distance. None of these properties is reproduced by figure 7.1, quite simply because the cusps do not represent the potential of the ligand itself, but the *effects* of that potential upon the metal wavefunction. Thus the heights of the cusps are not constant, but vary (inversely) with metal–ligand distance: that is, the effect of the point-charges decreases as they are removed from the metal. Similarly the second- and fourth-order cusps in figure 7.1 describe how the *effects* of the potential fall off with distance in second- and fourth-order. We could

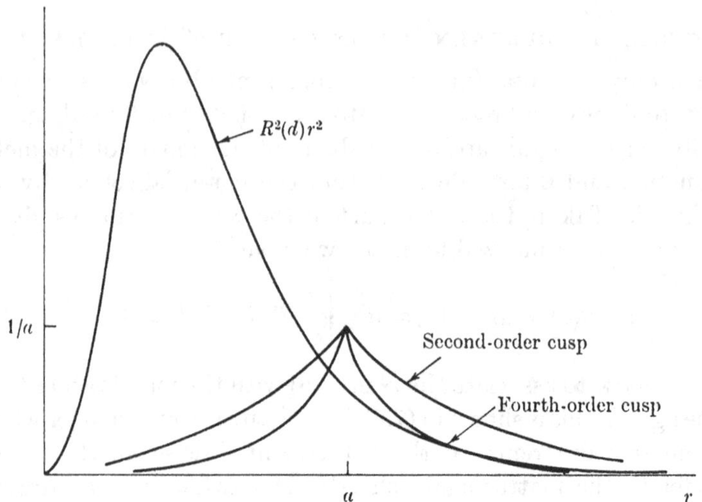

Figure 7.1. Breakdown of G^l functions.

have arranged for the cusp heights to remain constant with distance by choosing to associate the r^2 term of the differential increment in (7.5) partly with the wavefunction and partly with the potential, *viz.*

$$G^l = \int_0^a R^2(d)r.\frac{r^{l+1}}{a^{l+1}}.dr + \int_a^\infty R^2(d)r.\frac{a^{l+1}}{r^{l+1}}.dr. \tag{7.9}$$

Cusp heights are now normalized to unity regardless of the value of a, but the wavefunctions are normalized in an unusual way. Nothing is gained by this procedure: we mention it merely to emphasize the artificiality of separating the functions under the integral sign and to note how the cusps describe effects of potentials rather than potentials themselves. In passing, note here that the separation of G^l integrals into parts dependent on metal only and parts dependent on ligand only apparently reflects the factorizability property of the Spectrochemical Series: we cannot emphasize too strongly that such a conclusion is specious if only because of the arbitrary nature of the splitting up of functions under the integral sign in (7.6).

PROPERTIES OF THE RADIAL INTEGRALS

We now investigate properties of the radial integrals G^l. Our initial discussions are qualitative and can be appreciated mainly by reference to diagrams alone. The process is 'quantified' in an appendix to this chapter in which the intuitive conclusions of the earlier treatment are supported by algebraic and numerical evaluation of the G^l and related integrals, following Ballhausen and Ancmon.[4] However, we cannot emphasize too much that any calculation of radial parameters based on the point-charge or point-dipole models can only be taken as illustrative: a detailed mathematical analysis of what immediately follows cannot furnish 'proof'.

Our treatment takes the form of a series of discussions in which we attempt to relate experimental observations of radial parameters with generally applicable notions of bond length, charge transfer, the electroneutrality principle, polarizability and so on. We begin by establishing some general principles and then successively apply these to various pieces of experimental data. At first the arguments presented are simple and direct, but they gradually become more involved and our premises need to be re-examined in more detail. We persist with the approach in order to illustrate both successes and difficulties, but principally to sketch an approach to the understanding of radial parameters. Following this series of discussions we summarize the nature of the information and ideas gained and see what may be considered likely and what uncertain. In particular, we list areas of research in which the approach suggests more experiments are desirable.

BOND LENGTHS

The first point to be made on the subject of bond lengths in crystal-field theory, one which must always be borne in mind, is that the point-charge formalism must regard a as an *effective* bond length. Ligands obviously cannot behave as point-charges, but if we choose to so represent them, there can be no justification for treating bond lengths in any more absolute a way. Although they are not of immediate interest, bond angles must similarly be regarded as *effective*† parameters of the system as discussed in chapter 4.

In terms of the definitions in (7.6), as a increases, the potential cusps decrease in height: at the same time the amplitude of the wavefunction decreases. Both factors, which are really only one physical effect, contribute to decreasing G^l functions with increasing a. Therefore both Dq and Cp decrease with increasing bond length. But they do not decrease at the same rate. It is interesting to observe the behaviour of the ratio Cp/Dq with effective bond length. Note that

$$Cp/Dq = \tfrac{12}{7}.G^2/G^4, \tag{7.10}$$

† We are well aware that the word 'effective' is redundant in qualifying the word 'parameter' but we prefer to retain it here for reasons of emphasis.

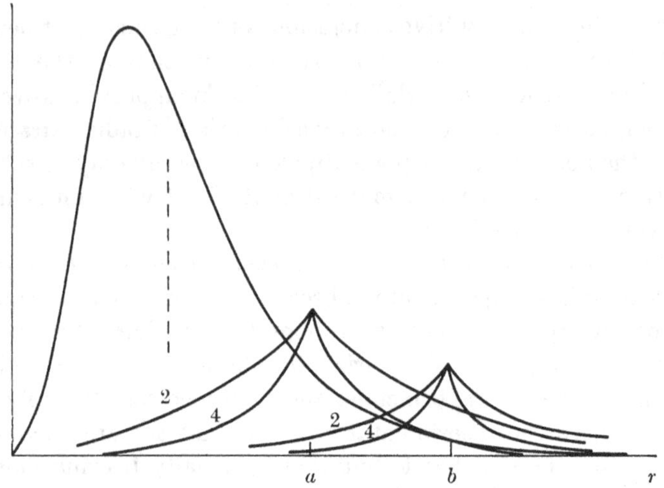

Figure 7.2. Components of G^l functions for two bond lengths.

based on the definitions of (7.7). In figure 7.2 we plot cusps for two different bond lengths. The main contribution to the G^l integrals comes from that region near the maximum in $R^2(d)r^2$ – near the dotted line in figure 7.2, say. This argument depends upon the relative disposition of the maxima in $R^2(d)r^2$ and in the potential cusps shown in figures 7.1 and 7.2. We assume this is the case here and gain support for the assumption from the more quantitative calculations in the appendix and comments therein: the assumption is important, however, as we shall see shortly. Now the ratio of the cusp terms in (7.6) of order two and four is:

$$\left.\begin{aligned} \rho_< &= \frac{r^2/a^3}{r^4/a^5} = \frac{a^2}{r^2} \quad \text{for} \quad r < a, \\ \rho_> &= \frac{a^2/r^3}{a^4/r^5} = \frac{r^2}{a^2} \quad \text{for} \quad r > a. \end{aligned}\right\} \tag{7.11}$$

This means that the ratio ρ of effective potentials between second- and fourth-order is always greater than or equal to one and increases with distance from the cusp. The region of maximum importance so far as the total integrals G^l are concerned is further from the cusp for the greater bond length b in figure 7.2 than for the lesser a, and so for the total integrals we expect:

$$G^2/G^4 \text{ increases as bond length increases.} \tag{7.12}$$

We can also enquire how sensitive this trend is to bond length. While this is done more formally in the appendix, the following qualitative discussion may be useful.

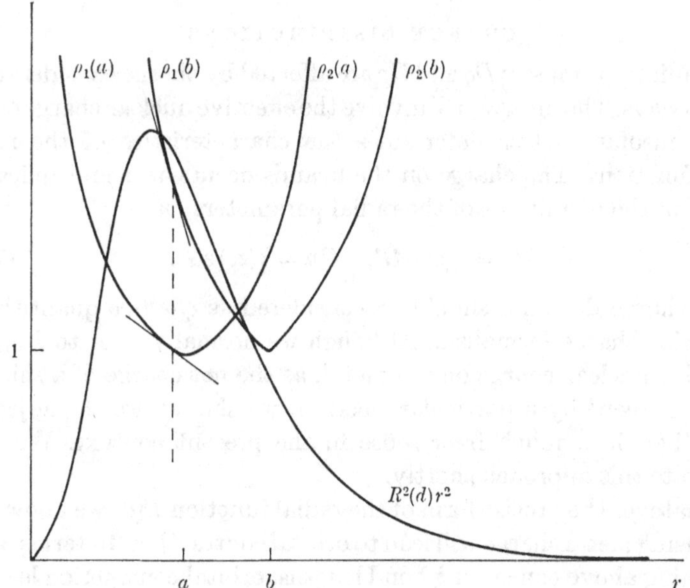

Figure 7.3. To illustrate variation of sensitivity of G^2/G^4 ratios with bond length.

In figure 7.3 we represent the effective potential ratios, ρ of (7.11), corresponding to the situation in figure 7.2. By differentiating ρ in (7.11) with respect to r we have the slope of the ratio of potentials at distance r from the origin as:

$$\left.\begin{aligned}\frac{\partial\rho_1}{\partial r} &= -\frac{2a^2}{r^3} \quad \text{for ligand at } a \text{ and } r < a,\\ \frac{\partial\rho_1}{\partial r} &= -\frac{2b^2}{r^3} \quad \text{for ligand at } b \text{ and } r < b.\end{aligned}\right\} \tag{7.13}$$

Therefore, at a given distance from the metal, the slopes of the two potential ratios for the two ligands distant a and b are in the ratio a^2/b^2. Again, let the dotted line represent the region where contributions to the G^l function (not their ratios) are maximized. We see from figure 7.3 that the sensitivity of the ratio G^2/G^4 to bond length change is less for short bond lengths than for long ones. That is, a small change in a larger bond length will alter the Cp/Dq ratio more rapidly than a similar change in a shorter bond. On the other hand, by comparison of the slopes of the cusps in figure 7.2, we see that absolute changes in Dq or Cp will be more sensitive to changes in bond length when bonds are short.

CHARGE DISTRIBUTIONS

The radial parameters Dq and Cp are affected by charge considerations in two ways. The integrals G^l involve the effective nuclear charge on the metal insofar as this determines the characteristics of the radial wavefunctions. The charge on the ligands occurs as a multiplicative factor in the definitions of the radial parameters, *viz.*

$$Dq = \tfrac{1}{6}(ze)\,eG^4, \quad Cp = \tfrac{2}{7}(ze)\,eG^2. \tag{7.14}$$

Both charge densities should be considered as *effective* quantities in the point-charge formalism. Although we normally refer to Z_{eff}, the effective nuclear charge on the metal, as the net charge of a shielded nucleus 'seen' by a particular electron, we should use the adjective 'effective' in a much freer sense in the present context. We shall illustrate this approach shortly.

Whatever the precise form of the radial function $R(d)$ we know that increasing metal charge will lead to orbital contraction. In terms of the reasoning above concerning bond lengths, orbital contracticn leads to the same qualitative result as bond lengthening. In particular, we get the relation:

$$G^2/G^4 \text{ increases as } Z_{\text{eff}} \text{ increases.} \tag{7.15}$$

We may envisage a change in Z_{eff} as due to a transfer of charge between ligands and the central metal. In this sense, covalency effects are empirically incorporated into the crystal-field model. A ligand donating negative charge to the metal, decreases Z_{eff} on the metal, expands the 'non-bonding' d orbitals which then repulsively interfere more with the 'point-charges': in consequence all radial parameters – Dq and Cp – increase, but Dq proportionately more so because G^2/G^4 decreases.

We must remember, however, that the effects of bond shortening and orbital expansion need not be identical in detail. The relative rates of change of Dq and Cp/Dq due to one effect need not be the same as for the other. The point-charge formalism requires us to talk qualitatively on such matters so we shall not attempt to quantify these differences: it is sufficient for the moment to note that they will exist, leaving further discussion till later.

Now a transfer of charge from the ligands to the metal requires a loss of charge on the ligands. This is represented by a decrease in the factor ze in (7.14) by which the radial integrals are multiplied. In this way, a decrease in ligand charge ze reduces both Dq and Cp, *by the same proportion.*

SUMMARY OF TRENDS

Summarizing, we have three quantities which may affect radial parameters – bond length, metal charge and ligand charge – and in the point-charge model all three must be seen as *effective* rather than absolute quantities. Variation in these quantities affects Dq and the ratio Cp/Dq according to the three principles of table 7.1. Effects I and II are more sensitive for Dq at shorter bond lengths and for Cp/Dq at longer bond lengths. However, and this is most important, as effects I and II are equivalent in their qualitative results, if not their origins, the meanings of 'shorter' and 'longer' as descriptions of bond lengths depend on the relative roles these two mechanisms play. This is discussed more later.

TABLE 7.1

I	Increasing bond length	Dq decreases, Cp/Dq increases
II	Transfer of negative charge to metal (i.e. decrease Z_{eff})	Dq increases, Cp/Dq decreases
III	Transfer negative charge from ligand (i.e. decrease ze)	Dq decreases, Cp/Dq is unchanged

We now use these trends in considerations of the various experimental aspects of radial parameters listed at the head of this chapter, and before.

RADIAL PARAMETERS AND COORDINATION NUMBER

Currently available experimental data suggest that increase in coordination number has little effect on Dq but serves to increase the ratio Cp/Dq significantly. Let us consider tetrahedral and octahedral species MX_4 and MX_6. Other things being equal, we can expect longer bonds in an octahedron than in a tetrahedron so that factor I above predicts lower Dq values and higher Cp/Dq ratios in MX_6. Remember that we are referring to Dq as defined in (7.14) and not to Δ_{oct} and Δ_{tet}: thus the 'geometric factor' $-\frac{4}{9}$ has been allowed for already. The invariance of Dq with coordination number or, put another way, the fact that $\Delta_{\text{tet}} \sim -\frac{4}{9}\Delta_{\text{oct}}$ holds fairly well in practice, is quite surprising and may be seen as one more of Nature's coincidences.

The Nephelauxetic effect (see chapter 9) is slightly greater in tetrahedral ions than in octahedral: this is commonly interpreted to imply a smaller Z_{eff} in the tetrahedron, so far as the 'spectral' *d* orbitals are concerned, but only to a slight extent. An argument[78] put forward in favour of smaller Z_{eff} values in tetrahedra involves assumed sp^3 hybrid bonding orbitals in the tetrahedron and d^2sp^3 orbitals in the octahedron. Charge transferred to the metal *via* such orbitals would attain more *s* character in the tetrahedron than in the octahedron. The greater penetration of the *s* functions would then shield the metal *d* orbitals more in the tetrahedron with resultant 'cloud expansion'. Whatever the detailed explanation, if there is a decreased Z_{eff} on the metal, then factor II above predicts larger Dq values and smaller Cp/Dq ratios in the tetrahedron than in the octahedron. In any case the effect is likely to be small.

A third factor comes from the application of the electroneutrality principle. In order to achieve somewhat similar (and small) net charges on the tetrahedrally- and octahedrally-coordinated metal atoms, four ligands must donate about the same charge as six. Thus each ligand in the tetrahedron must donate roughly† $\frac{6}{4}$ times the charge transferred from each ligand in the octahedron. Accordingly, on changing from four to six coordination, other things being equal, ze per ligand increases so that, according to factor III above, Dq increases and Cp/Dq remains unchanged.

Thus, increasing the coordination number from four to six should give rise to the following trends:

I	Bond lengths increase	Dq decreases, Cp/Dq increases;
[II	Z_{eff} increases?	Dq decreases, Cp/Dq increases];
III	ze increases	Dq increases, Cp/Dq is unchanged.

There is ample experimental evidence to show that Dq values change little with coordination number, so that factors I and II must be balanced by factor III so far as Dq is concerned. However this balance is achieved, and we guess that variation in factor II is not particularly important here, there will be no such cancellation with respect to Cp/Dq ratios. We have the unambiguous prediction that Cp/Dq ratios will increase with increasing coordination number.

In the foregoing argument we have hopefully avoided the common trap of reaching a conclusion which depends upon approximate cancellations between values of two or more quantities which, by the nature of the model, are little more than educated guesses. Thus, rather than assert that the model predicts little variation of Dq with

† 'Roughly' because Pauling's criterion accepts variation of $\pm\frac{1}{2}$ unit charge on the metal.

coordination number, we are content to see how a given *Dq* value may be achieved and examine what the consequent behaviour of *Cp/Dq* must be.

INVERSE TRENDS IN *Dq* AND *Cp/Dq*: EFFECTIVE BOND LENGTHS

The most general trend to emerge from our treatment of experimental data in chapters 4 and 5 was that *Cp/Dq* increases along a series as *Dq* decreases, for larger *Cp/Dq* ratios. Theoretically this inverse behaviour follows immediately from factors I and II in table 7.1, considered separately. Before looking at this in detail, however, we need to consider the case of an exception to the general rule. We observed in chapter 4 that both *Dq* and *Cp/Dq* decrease in the tetrahedral CuX_4^{2-} ions as X = Cl is replaced by X = Br. Perhaps the accuracy was such that it is best to conclude that while *Dq* decreases, *Cp/Dq* changes very little.

The Cu–Br bond lengths as determined by X-ray methods are longer than the Cu–Cl and we should presumably carry this over to the effective bond lengths as represented in the point-charge model. Mechanism I of table 7.1 predicts a lower *Dq* value for the tetrahedral bromide which is observed, but at a larger *Cp/Dq* ratio relative to the chloride, which is not. The bromo-complex is supposed to be more covalent in the sense of a greater transfer of negative charge from the bromines than from the chlorines. Such a conclusion is based, for example, on spectral intensities, on orbital reduction factors, and on the instability of cupric iodide (assumed to transfer even more charge to the metal) with respect to cuprous iodide. The effective nuclear charge on the metal should be less in $CuBr_4^{2-}$ than in $CuCl_4^{2-}$ and so expand the 'non-bonding' *d* orbitals: this is also consistent with the greater Nephelauxetic effect in tetrahedral bromides compared with chlorides. Factor II then predicts $Dq(\mathrm{Br}) > Dq(\mathrm{Cl})$ and *Cp/Dq* smaller for the bromide. Therefore we have opposed trends from factors I and II. Factor III, involving lower charge on the bromines than on the chlorines, favours $Dq(\mathrm{Br}) < Dq(\mathrm{Cl})$ without affecting the ratios *Cp/Dq*. If the qualitative ideas of the present treatment are to be consistent with experiment, then factors I and II must be assumed to cancel approximately, perhaps with II slightly more important. Thus, on replacing chlorine by bromine in the CuX_4^{2-} ions, *Cp/Dq* changes little, perhaps decreasing if factor II is more important. The charge on the ligands will not affect this conclusion. The assumed roles of factors I and II leads to little change or a slightly higher *Dq* value for the bromide. Factor III finally reverses this trend, so reproducing the experimental observations. The fact that Cp/Dq ratios are small (1–2) means we are to regard the effective Cu–X bonds as short. This conclusion is consistent with the trends underlying factors I and II. The actual bonds are fairly short (at least, compared with their octahedral counterparts) and are effectively made more so by the expansion of the metal wavefunctions resulting from a transfer of charge from the ligands.

The preceding discussion was made to establish a *modus operandi*. The discussion emphasizes the idea that bond lengths should be treated as *effective* quantities in this crystal-field model. This concept comes not only from the philosophical point that all variables in the point-charge model should be treated in this way, but also from the inseparability of factors I and II. The consequences of a shorter bond are similar to those of a decreased Z_{eff} on the metal. This means, of course, that predictions of the relative sizes of Dq values are not usually possible, at least if bond length and charge transfer trends are opposed, but the interpretation of experimental trends in Dq values along these lines should allow predictions of Cp/Dq ratios. In short, we are hoping for self-consistency and clarification rather than a general prediction of Dq values.

The Dq values in $CuCl_4^{2-}$ and $NiCl_4^{2-}$ are about 1150 and 730 cm^{-1}, respectively. A consideration of shielding effects would suggest a greater effective nuclear charge for Cu^{2+} than for Ni^{2+}, and hence more contracted orbitals for Cu^{2+}, and so favour the reverse trend in Dq values. On the other hand, charge transfer and covalency are probably greater in tetrahedral copper(II) complexes than in tetrahedral nickel-(II) ones and we must assume that this effect overtakes the greater free-ion shielding. A lesser importance of factor II in $NiCl_4^{2-}$ leaves Dq smaller in $CuCl_2^{2-}$ and at the same time Cp/Dq larger. This was the trend deduced from the spectral work on these ions described in chapter 4. It seems that the greater covalency in the copper systems makes for a shorter effective bond so that both Dq is larger than otherwise expected and Cp/Dq smaller. Such an agreement must appear tendentious but it does at least provide a systemization of our knowledge and suggest further experiments to check the broad conclusions.

It is in this spirit that we would 'interpret' the observation that Dq increases with oxidation state in terms of the increased polarizing power of the formally higher charged metal ions, rather than concentrating solely on the decrease in orbital size. What we are arguing for, in effect, is a modification of the simple point-charge repulsion picture by the broad principles of electroneutrality, polarizability and 'intuitive molecular-orbital theory': this is commonly called *ligand-field theory*. By adopting this approach we may retain the simplicity of the point-charge model in relating so many spectral and magnetic properties. While the greater *Dq* values for metals in the second- and third-row of the transition block might seem to reflect the greater extension of 4*d* and 5*d* orbitals, their greater diffuseness suggests that we should be content to accept the observed *Dq* values and use the foregoing ideas to predict and interpret *Cp*/*Dq* values. There appears, however, to be insufficient low-symmetry data on these systems available at present.

We return to the inverse trends of *Dq* and *Cp*/*Dq* discussed in earlier chapters. These were mostly based on molecules of the type *trans*-$M(base)_4X_2$ as discussed in chapter 5. The larger *Dq*(X) is, the smaller *Cp*/*Dq* appears to be, though for these D_{4h} symmetry molecules, no absolute values for *Cp*/*Dq* can be derived. It appears that we must concentrate on the bond length factor (I) in discussing these observations. If *Dq* variations are determined primarily by this factor (as is obviously likely for the $Nipy_4Cl_2$ and $Nipyrazole_4Cl_2$ molecules – *q.v.*) then larger bond lengths imply smaller *Dq* values and larger *Cp*/*Dq* ratios and the picture is very simple. However, if factor II were significant, a change of Z_{eff} should also affect the equatorial *Dq* values and, to a first approximation at least, this does not appear to be the case. Insofar as this approximation is valid, it thus appears that the requirements of the electroneutrality principle are mainly satisfied by the four equatorial bases. This picture represents an extreme, however, and it may well be worth experimentally examining *Dq* values in these D_{4h} complexes in more detail.

Another aspect of the problem must be associated with the larger bond lengths actually present in octahedral complexes. Relative to the tetrahedral molecules we have discussed, the *effective* bond lengths must be larger also, in which case, not only are *Cp*/*Dq* values larger, but they are also more sensitive to small changes in bond length as summarized above table 7.1. Further, as shown in chapter 5, some of the axial *Dq* values are abnormally low (i.e. with respect to values from the Spectrochemical Series) and almost certainly indicate steric interaction of the axial ligands with the equatorial bases. For all these reasons, therefore, we conclude that any inverse trends in *Dq* and *Cp*/*Dq* in D_{4h}-symmetry molecules predominantly reflect the consequence of effective bond length change. Here again, it would be interesting to have more reliable low-symmetry data on such molecules suffering the reverse sense of tetragonality: and to gain knowledge of some absolute *Cp*/*Dq* ratios, upon which the inverse trends rely.

FACTORIZABILITY OF THE SPECTROCHEMICAL SERIES

A most significant feature of the Spectrochemical Series is the way *Dq* values may be expressed as products of functions involving ligand only and metal only – the *f* and *g* terms listed in chapter 1. It is important to note that the factorizability is only approximate, but nevertheless it is still remarkable that the relation holds at all. At first sight it seems difficult to reconcile with the trends in table 7.1 for we appear there to have factors not only related to the metal or to the ligands but also to both. Thus our notion of 'effective' bond lengths depends in part on the way charge is shared by ligand and metal. As with the discussion of inverse trends in *Dq* and *Cp*/*Dq* the assumption that factor I is more important than factor II in determining changes in octahedral *Dq* values (and we note that Spectrochemical *Dq* values are mostly derived from octahedral molecules) can help. On this basis, the *f*.*g* factorizability would follow from the additivity of covalent radii which, despite several 'exceptions' is a fairly good rule in structural chemistry. In some qualitative way, one can also see how metal polarizing power and ligand polarizability (factor II minimized) are properties independent of metal-and-ligand together and do contribute to the factorizability of *Dq*. It does not seem possible to offer an 'explanation' of the factorizability relationship more concrete than the previous remarks, as any discussion on this subject must inevitably rest on assessments of the magnitudes of the various effects rather than simply on trends. Nor does it seem obvious to us that the *f*.*g* factorizability for *Dq* should carry over for *Cp* or the ratio *Cp*/*Dq*. This clearly defines another area of current interest involving the determination of many more *Cp* values from spectral and magnetic properties.

SOME DIFFICULTIES

Within the scope of our current approach to the properties of crystal-field radial integrals we appear to have come to the position of supposing changes in Z_{eff} along a series of octahedral molecules to be relatively unimportant. At least if this were so many of the experimental observations would follow more directly. Of course, to some extent small changes in Z_{eff} can be accommodated within the concept of changes in effective bond length as discussed above. Inevitably in qualitative theories like this, the greatest difficulties occur when one needs to assess the relative magnitudes of opposing effects. Probably the best one can do is acquire some 'feel' for these quantities from comparison of data throughout as wide a field as possible and this we have tried to do, given the very short supply of low-symmetry ligand-field parameter values.

A similar problem arises when we consider trends in some *absolute Cp* values rather than *Cp*/*Dq*. The values for the $NiCl_4^{2-}$ and $CuCl_4^{2-}$ ions furnish an interesting example on this theme. The discussion developed earlier rationalized the greater *Dq* value for the cupric ion relative to the nickel ion in terms of an important role for factor II and deduced that this would be accompanied by a decrease in *Cp*/*Dq*. This is observed experimentally. However, consideration of factors I and II *separately*, shows that while an increase in *Dq* would be accompanied by the decrease in *Cp*/*Dq*, the absolute value of *Cp* should increase with *Dq*. *Dq* values for $NiCl_4^{2-}$ and $CuCl_4^{2-}$ are *ca.* 730 and 1150 cm^{-1}: *Cp*/*Dq* ratios were, respectively, *ca.* 6.5 and *ca.* 2. Thus, if the experimental data are to be believed, and some doubts were expressed in chapter 4, then as *Dq* increases from 730 to 1150 cm^{-1}, *Cp decreases* from *ca.* 4600 to *ca.* 2100 cm^{-1}. According to our reasoning earlier, neither factor I nor II, taken separately, could explain this behaviour.

Within the present framework, these results appear explicable only if the opposing trends represented by factors I and II are *quantitatively* different. That is, we must suppose that the difference between the rates of change of *Dq* with respect to factors I and II are not the same as the corresponding behaviour of *Cp* with respect to these two factors. It is obviously idle to guess how these behaviours will differ: the more quantitative treatment in the appendix lends some support to these ideas. It may well be that tetrahedral copper(II) ions share an atypical slot in the scheme of things, in which variation in factor II is relatively more important than elsewhere in the series. Once more this can only be assessed properly when we have collected far more data throughout the whole transition block, particularly from the extreme left of the series and from the second and third rows.

A PICTORIAL REPRESENTATION OF *Ds* AND *Dt* VALUES: THE 'STRENGTH' OF A LIGAND FIELD

It is often helpful to draw diagrams to aid conception or to act as pictorial mnemonics. Let us represent *Dq* and *Cp* values in a tetragonally-distorted octahedron by ellipsoids of revolution whose half-axis lengths are proportional to the magnitudes of these quantities. In figure 7.4 we show two examples. This pictorial representation of *Dq* and *Cp* values embodies ideas of *Cp* being generally larger than *Dq*, and that larger *Dq* values are usually accompanied by relatively smaller

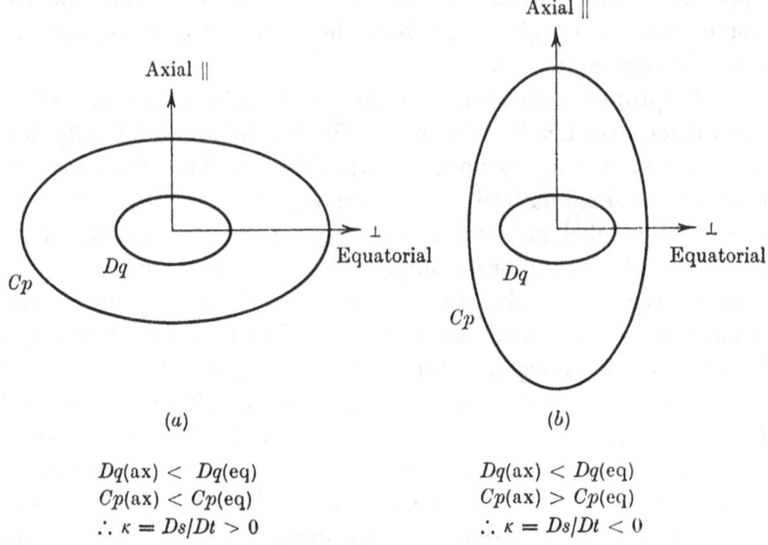

Figure 7.4. A pictorial mnemonic for *Dt* and *Ds* (*a*) κ positive, (*b*) κ negative.

Cp/*Dq* ratios, though not necessarily by smaller absolute *Cp* values. In the case of figure 7.4(*b*), the reversed sense of anisotropy of *Cp* relative to *Dq* implies a negative κ value. The inverse trend in *Cp*/*Dq* relative to *Dq* would not have emerged from the data in chapter 5 had there been large values of *Ds* of an appropriate sign, as discussed at the end of that chapter. All this should be clear from the diagrams in figure 7.4.

The representation also emphasizes the relative magnitudes of *Cp* with respect to *Dq* and this fact raises an interesting point on the general topic of ligand fields. It is common to refer to the 'strength' of a crystal-field, meaning the magnitude of 10*Dq*. One might suppose that a strong field is strong for all orders of perturbation but, as the summary in figure 7.4 shows, this need not be so. The situation in figure 7.4(*b*) corresponds to a stronger field in the equatorial plane as defined by *Dq*, but stronger parallel to the unique molecular axis as defined by *Cp*. It is really an accident of symmetry, as it were, that the effects of *Cp* vanish identically in cubic symmetry, so historically focussing attention on *Dq* instead.

The point is this: we tend to regard distortions from cubic symmetry as small 'corrections' and so easily neglected. While it is true that energy levels and wavefunctions only alter with respect to *Dq* in

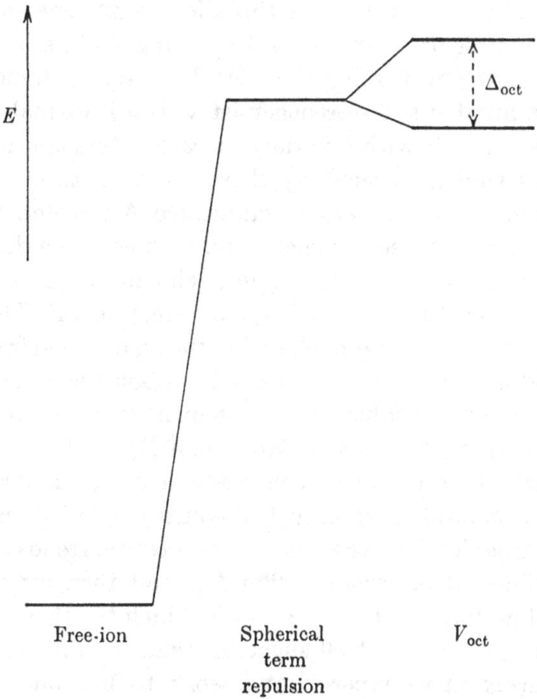

Figure 7.5. 'Crystal-field strengths'.

cubic symmetry because such symmetry causes second-order terms to vanish, such alterations which do occur on distortion may do so very rapidly with respect to small geometric displacements as a result of the large 'coefficients of distortion'; that is, the second-order radial integrals. The relative size of Cp compared with Dq needs occasion no real surprise if we recall the sort of diagram frequently used when first teaching the elements of crystal-field theory. Thus in figure 7.5, we represent a general 'spherical term' of repulsion (from Y_0^0 in the potential) as a large destabilization. The splitting of the d orbitals by the fourth-order terms in V_{oct} (say) is small in comparison. In terms of this diagram, the magnitudes of radial parameters associated with zeroth-, second- and fourth-order terms in the potential, decrease in that order. Calculations in the appendix support this assertion.

SUMMARY

There are obviously great objections to a point-charge model for radial parameters, many of which were enumerated in our first chapter. Equally there are great conveniences in retaining the point-charge formalism at least because this allows extensive and relatively simple correlation between spectral splitting factors and magnetic susceptibilities of a wide range of molecules, varying in metal, ligand, coordination number and stereochemistry. But if we make use of this simplicity we are left with a variety of parameters and lists of parameter values which, particularly those referring to low symmetry, have little immediately apparent significance. A first step, therefore, is to collect values for these parameters under a common definition and then to see if any semi-empirical approach can rationalize them and indicate which are 'normal' and which 'exceptional'. The modification of the point-charge approach within the ligand-field framework, in which ligand and metal charges as well as bond lengths and angles become 'effective' variables rather than absolute ones constitutes such a semi-empirical approach. After all, if Dq itself may be regarded as a *parameter*, then its constituent parts may be similarly regarded (though not separately determined, of course). The approach appears to be most valuable when we adjust the theoretical trends to reproduce some experimental observables like Dq, and then infer associated trends in other data like Cp. The way in which the theoretical trends must be manipulated, in itself suggests trends and indicates areas of doubt requiring more experimental work to be done. In the immediately preceding section we discussed ideas about the 'strengths' of ligand-fields in different orders. It should be clear that the magnitude of the second-order parameter Cp is important for distorted systems and this has relevance, for example, for those interested in crystal-field stabilization energies as applied to transition states in kinetic studies.

Obviously, it is desirable to accumulate very much more information on low-symmetry systems in general. There are particularly prominent gaps in our knowledge for the second- and third-row transition block elements, for D_{4h} symmetry molecules with the reversed sense of tetragonality to those discussed in chapter 5, and for Cp values of high-field, π-bonding ligands like CN^- etc. Following some of our earlier discussions, we would ask whether copper(II) ions are unusual, uniquely or otherwise, in that variations in radial parameters may not be determined mostly by effective bond length considerations as appears to be the case elsewhere. Finally we have the whole field of the lanthanides where a pure crystal-field approach might be expected to be more apposite: we give appropriate integrals in the appendix following.

APPENDIX 7A. EVALUATION OF ELECTROSTATIC RADIAL INTEGRALS

Ballhausen and Ancmon[4] have calculated G^l and other integrals using the point-charge model we have discussed in this chapter. Following the very early calculations of Van Vleck[113] and of Polder[85] restricted to Dq values alone, they represented the metal wavefunctions by hydrogenic functions. For metal $3d$ orbitals they used the radial function:

$$R(3d) = \frac{4}{81\sqrt{30}}\left(\frac{Z}{a_0}\right)^{\frac{3}{2}}\left(\frac{Zr}{a_0}\right)^2 e^{-Zr/3a_0}. \tag{7A.1}$$

This function is normalized to unity by the relation:

$$\int_0^\infty R^2(3d)\,r^2\,dr = 1, \tag{7A.2}$$

a_0 is the Bohr radius and $Z \equiv Z_{\text{eff}}$, the effective nuclear charge. We can assume that suitable values of Z_{eff} might be arrived at by using Slater's rules.[23, 104]

The radial integrals we want, therefore, are given by:

$$G^l = \int_0^a R^2(3d)\,.\frac{r^{l+2}}{a^{l+1}}\,dr + \int_a^\infty R^2(3d)\,\frac{a^l}{r^{l-1}}\,dr,$$

$$= \frac{4^2}{81^2.30}\left(\frac{Z}{a_0}\right)^7\left[\int_0^\infty r^4.\frac{r^{l+2}}{a^{l+1}}.e^{-2Zr/3a_0}dr\right.$$

$$\left.-\left\{\int_a^\infty r^4\frac{r^{l+2}}{a^{l+1}}\,e^{-2Zr/3a_0}\,dr - \int_a^\infty r^4\frac{a^l}{r^{l-1}}.e^{-2Zr/3a_0}\,dr\right\}\right]. \tag{7A.3}$$

Defining
$$y = 2aZ/3a_0$$

we have
$$\frac{2Zr}{3a_0} = y.\frac{r}{a}, \tag{7A.4}$$

and
$$G^l = \frac{4^2}{81^2.30}\left(\frac{Z}{a_0}\right)^7\left[\left(\frac{1}{a}\right)^{l+1}\int_0^\infty r^{l+6}.e^{-y.r/a}\,dr\right.$$

$$\left.-\left\{\left(\frac{1}{a}\right)^{l+1}\int_a^\infty r^{l+6}.e^{-y.r/a}\,dr - \left(\frac{1}{a}\right)^{-l}\int_a^\infty r^{5-l}.e^{-y.r/a}\,dr\right\}\right] \tag{7A.5}$$

Using the standard integrals:[23]

$$\int_0^\infty x^m e^{-cx}\,dx = \frac{m!}{c^{m+1}},$$

we have:
$$\left(\frac{1}{a}\right)^{l+1}\int_0^\infty r^{l+6}.e^{-yr/a}\,dr = \left(\frac{a}{y}\right)^6\frac{(l+6)!}{y^{l+1}}, \tag{7A.6}$$

and obtain:

$$G^l = \frac{Z}{1080a_0}\left[\frac{(l+6)!}{y^{l+1}} - y^6\left\{\left(\frac{1}{a}\right)^{l+7}\int_a^\infty r^{l+6}e^{-yr/a}dr\right.\right.$$
$$\left.\left.-\left(\frac{1}{a}\right)^{6-l}\int_a^\infty r^{5-l}.e^{-yr/a}.dr\right\}\right]. \quad (7\text{A}.7)$$

The remaining definite integrals are evaluated using *Tables of Molecular Integrals*,[64] as follows: Defining

$$A_n(\alpha) = \int_1^\infty e^{-\alpha\lambda}.\lambda^n.d\lambda, \quad (7\text{A}.8)$$

we have from ref. 64, p. 56, the recursion formula:

$$A_n(\alpha) = \frac{1}{\alpha}\left[e^{-\alpha} + nA_{n-1}(\alpha)\right], \quad (7\text{A}.9)$$

whence:
$$A_n(\alpha) = \frac{e^{-\alpha}}{\alpha}\left[1 + \frac{n}{\alpha} + \frac{n(n-1)}{\alpha^2} + \ldots + \frac{n!}{\alpha^n}\right]. \quad (7\text{A}.10)$$

Using this relation together with nothing that

$$\frac{1}{a^{l+7}}\int_a^\infty r^{l+6}e^{-yr/a}.dr = \int_1^\infty r^{l+6}.e^{-yr}dr, \quad (7\text{A}.11)$$

we have:
$$G^l = \frac{Z}{1080a_0}\left[\frac{(l+6)!}{y^{l+1}} - y^6\left\{A_{6+l}(y) - A_{5-l}(y)\right\}\right], \quad (7\text{A}.12)$$

which is Ballhausen and Ancmon's 'master formula'.

Point-charge and dipole calculations are closely related. A point-dipole may simply be represented as an incremental dipole. The corresponding integrals for this model are simply

$$B^l = \frac{d}{da}G^l. \quad (7\text{A}.13)$$

Now
$$\frac{d}{d\alpha}\int_1^\infty e^{-\alpha\lambda}.\lambda^n.d\lambda = \int_1^\infty \frac{d}{d\alpha}(e^{-\alpha\lambda}.\lambda^n)\,d\lambda$$
$$= -\int_1^\infty e^{-\alpha\lambda}.\lambda^{n+1}.d\lambda, \quad (7\text{A}.14)$$

i.e.
$$\frac{d}{d\alpha}(A_n(\alpha)) = -A_{n+1}(\alpha). \quad (7\text{A}.15)$$

Using (7A.15) and the definition (7A.13) we get:

$$B^l = \frac{Z^2}{1620a_0^2}\left[-\frac{(6+l)!(l+1)}{y^{l+2}} - 6y^5\left\{A_{6+l}(y) - A_{5-l}(y)\right\}\right.$$
$$\left.-y^6\left\{A_{6-l}(y) - A_{7+l}(y)\right\}\right], \quad (7\text{A}.16)$$

which was also used by Ballhausen and Ancmon. Similar 'master formulae' for $4f$ electrons, appropriate for calculation of radial integrals in lanthanide complexes, derived as in ref. 40, are as follows:

$$G^l(4f,4f) = \frac{1}{48^2.35}\left(\frac{Z}{a_0}\right)\left[\frac{(l+8)!}{y^{l+1}} - y^8\left\{A_{l+8}(y) - A_{7-l}(y)\right\}\right], \quad (7\text{A}.17)$$

$$B^l(4f,4f) = \frac{1}{48^2.70}\left(\frac{Z}{a_0}\right)^2\left[-\frac{(l+1)(l+8)!}{y^{l+2}} - 8y^7\left\{A_{l+8}(y) - A_{7-l}(y)\right\}\right.$$
$$\left. + y^8\left\{A_{l+9}(y) - A_{8-l}(y)\right\}\right], \quad (7\text{A}.18)$$

where $y = Za/2a_0$. The hydrogen $4f$ radial wavefunction was taken as:

$$R(4f) = \frac{1}{768\sqrt{35}}\left(\frac{Z}{a_0}\right)^{\frac{9}{2}}.r^3.e^{-Zr/4a_0}. \quad (7\text{A}.19)$$

For calculational purposes we can estimate Z_{eff} from Slater's rules.†[23,104] The appropriate rule for d or f electrons is as follows. To find the shielding constant s, count 1 for each electron of an inside shell, 0 for those in an outer shell and 0.35 for each electron in the same shell. Z_{eff} is then given as the difference between the atomic number and the shielding constant. For ions in the first row of the d block we then find that Z_{eff} ranges from 3.65 for Ti^{2+} to 7.85 for Cu^{2+}. Ballhausen and Ancmon's tables of G^l and B^l integrals have been compiled for Z_{eff} varying in this range, together with bond lengths in the range 3.40 to 4.20 a.u. We shall not reproduce these tables here but rather use the same model to put some of our earlier discussions and diagrams on a more quantitative basis.

In figure 7A.1 we plot $R^2(3d)r^2$ for the $3d$ hydrogenic radial wavefunctions (7A.1) for three different effective nuclear charges. In the same figure are shown potential cusps P^l defined as

$$P^l = \frac{r^l_<}{r^{l+1}_>} \quad (7\text{A}.20)$$

for second- and fourth-order terms, corresponding to three different bond lengths. We have plotted these quantities to demonstrate the relative magnitudes and positions of these functions upon which we

† As discussed in chapter 6, Slater functions are over-expanded compared with Hartree–Fock functions. In particular, probability maxima are progressively overestimated from Slater's rules as the principal quantum number increases. Accordingly one must not pay a too-literal attention to the values of Z_{eff} used in these calculations. The gross assumptions of the model described at the beginning of the chapter remove any reason for using more refined forms of radial wavefunction anyway.

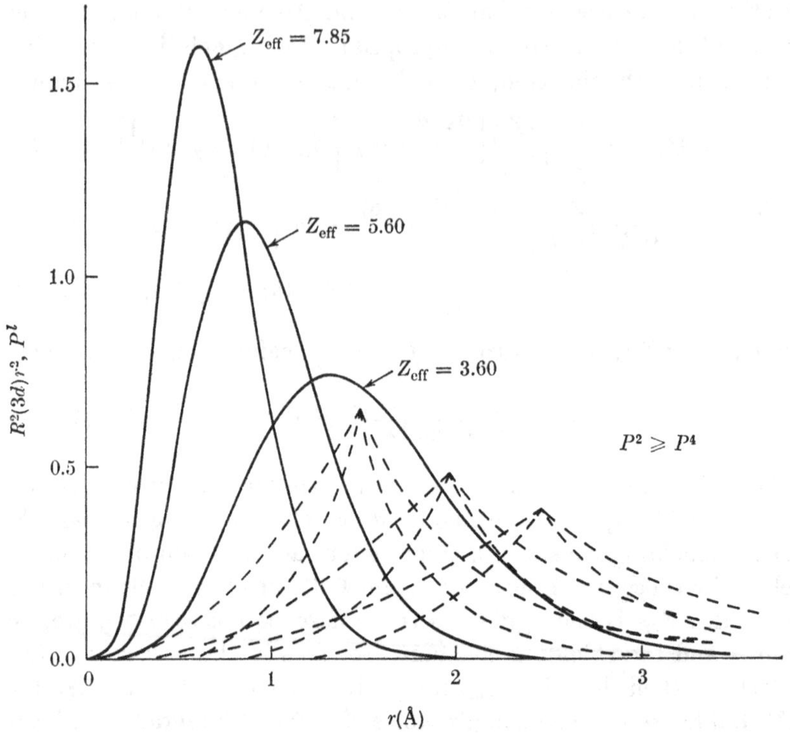

Figure 7A.1. Radial functions and cusps as functions of metal–ligand distance.

based the earlier figures 7.1 and 7.2. Another interesting representation is afforded by figure 7A.2 in which we plot the total product function under the integral sign in the G^l function. The curves in figure 7A.2 show the bond length regions where contributions to the total integrals are maximal. They also show how the relative magnitudes of G^2 and G^4 integrals are frequently determined more by regions closer to the metal: that is, maximal contributions to G^2 tend to occur nearer to the metal than for G^4. This appears to happen more for longer bond lengths and more expanded radial wavefunctions. Incidentally, this lends some quantitative support to the idea that the effects of shortening bond lengths or expanding metal wavefunctions are not equivalent in detail, as discussed in connection with our analysis of tetrahedral copper(II) and nickel(II) ions.

In our more qualitative discussions of the sensitivity of the ratio Cp/Dq to bond length, using figure 7.3, we concluded that G^2/G^4 ratios would increase with increasing bond length more sensitively for

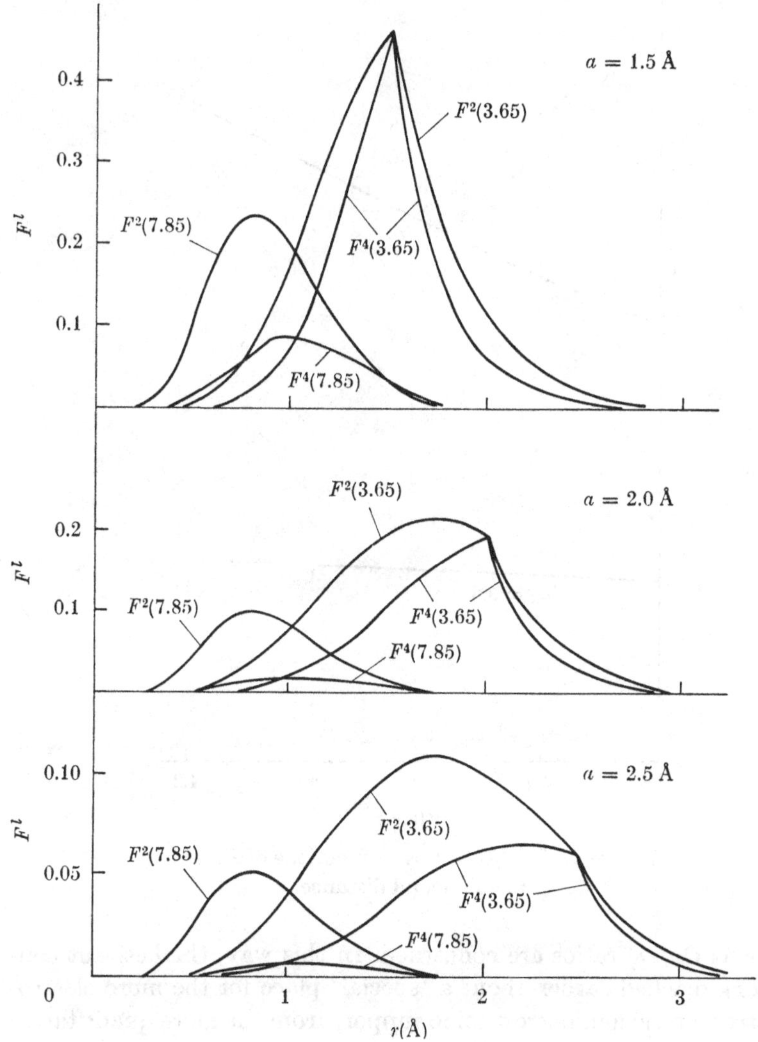

Figure 7A.2. Products of radial functions and cusps before integration to give G^l values.

longer bond lengths. Figure 7A.3 lends some support to this but again shows the different roles of factors I and II. Thus for small values of Z_{eff} the ratio Cp/Dq does not vary much with bond length. Reviewing our discussion on tetrahedral copper(II) ions again, it is possible that the extensive covalency in these ions, involving considerable transfer of negative charge to the metal, tends to reduce the effects of factor I

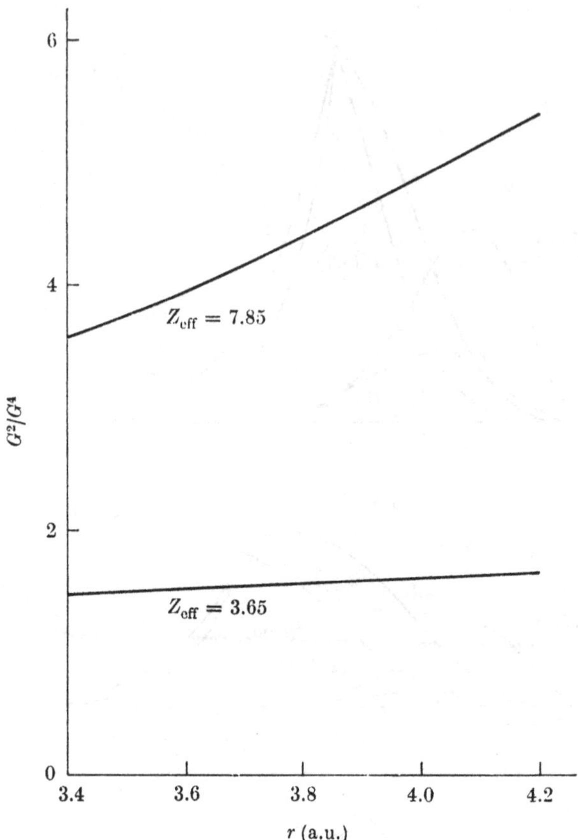

Figure 7A.3. G^2/G^4 ratios as functions of Z_{eff} and metal–ligand distance.

so far as Cp/Dq ratios are concerned. In this way, the hesitant conclusions reached earlier about a 'special' place for the more electronegative cupric ion, borrow some support from the more quantitative treatment.

8
SEMI-EMPIRICAL MOLECULAR-ORBITAL APPROACHES

It is the business of *ab initio* calculations to determine the energies and forms of the many-electron states which characterize the bonding in molecules. Approximate and semi-empirical calculations usually have somewhat more modest goals. The complexity of the *ab initio* calculations precludes their application to the vast number of co-ordination complexes that exist or will exist. There is clearly a real need for relatively simple theoretical models with which the chemist can investigate these complexes on a broad front; for it is with the trends within often closely related systems that major chemical interest lies. Qualitative and semi-quantitative features of chemistry are often not apparent from the purely numerical solutions of exact *ab initio* calculations while they are frequently apparent from 'chemical intuition'. Accordingly, approximate molecular-orbital methods are sought which provide theoretical validity coupled with conceptual utility.

An acceptable balance between these two sometimes conflicting aims of validity and utility is achieved by recourse to either empirical or semi-empirical methods on the one hand and approximate or semi-quantitative methods on the other. The former group seek to approximate many of the integrals of a molecular-orbital calculation by experimental observables. The Wolfsberg–Helmholz model provides the best-known example in coordination chemistry. Semi-quantitative methods, however, employ approximations within the theoretical framework of the Hartree–Fock formalism, the exact forms of these approximations being based upon reasonable physical assumptions. Included in this group are the 'zero-differential overlap' (ZDO)[80, 86] and the 'complete neglect of differential overlap' (CNDO)[87] methods,† which although originally set up to examine organic molecules have lately found application within the inorganic sphere. Mathematically, an important difference between these two approaches is the explicit use of a Hamiltonian in approximate techniques compared with a more implicit Hamiltonian in the semi-empirical models. One is never

† To some extent ZDO and CNDO calculations have also incorporated some empirical data.

quite sure which theoretical quantities are represented by the use of experimental observables in the semi-empirical methods. A discussion of this, and other matters relevant to approximate methods in transition metal chemistry, is given in a review by Dahl and Ballhausen.[a] In the present chapter we are concerned only with empirical and semi-empirical molecular-orbital models, most particularly with the angular overlap model of Schäffer and Jørgensen.[95]

The molecular-orbital formalism can be employed within a purely *empirical* model to rationalize the different bonding capacities of various ligands. A simple approach along these lines is one due to McClure.[70] The model, which is applicable to substituted octahedral complexes, considers the splittings of the e_g^* and t_{2g}^* orbitals to arise from the σ- and π-antibonding contributions of different ligands. Assuming that the symmetry of the molecule is such that the e_g^* orbitals are σ-antibonding only and the t_{2g}^* orbitals π-antibonding only, the orbital splittings are defined as functions of $d\sigma$ and $d\pi$. For example, in a *trans* di-substituted octahedron MA_4B_2 we have the situation shown in figure 8.1. The empirical parameters $d\sigma$ and $d\pi$ are given by:

$$d\sigma = \sigma_B - \sigma_A, \tag{8.1}$$

$$d\pi = \pi_B - \pi_A. \tag{8.2}$$

So comparing say $M(NH_3)_6$ with $M(NH_3)_4Cl_2$ we note that since ammonia is a stronger base then chlorine $\sigma_A > \sigma_B$ giving $d\sigma$ as negative. Further, as ammonia has no π-bonding effect, $\pi_A = 0$ so that $d\pi$ is positive. The predictions of the model can be compared with the experimentally observed band splittings and values for the two parameters obtained. Correlations between the parameter values and the bonding can be difficult if the axial ligands have both filled and empty π orbitals available for bonding since the sign of $d\pi$ will depend on which orbitals are used as well as their bonding strength relative to π-bonding equatorial ligands. Thus, Lever[67] quotes the value of $d\pi$ for $Ni(NH_3)_4(NCS)_2$ as being consistent with the interaction of the filled orbitals of the NCS^- ligand and the metal orbitals. This illustrates a general point made earlier that the energy level splittings depend upon several quantitative factors and so the 'sign of distortion' as reflected by the parameter values need not refer only to the distortion in a geometrical sense. This simple model has limited appeal and we do not pursue it further, turning our attention to the area of semi-empirical methods.

Essentially our approach is to sketch the background of the angular

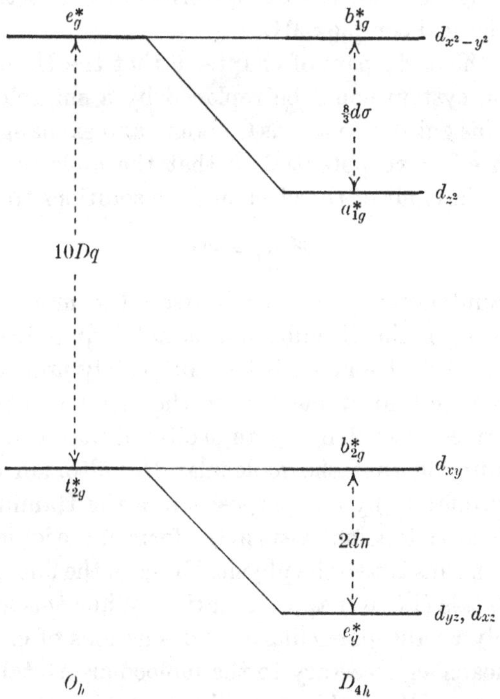

Figure 8.1. Definitions for McClure's empirical MO model.

overlap model introducing mathematical material as we require it, leaving somewhat more formal treatments of group overlap integrals, for example, as material for appendices. The angular overlap model is then discussed in some detail together with several recent examples of its use from the chemical literature. Finally, we make a comparison of the philosophies of the angular overlap method and the electrostatic crystal-field approach discussed in the previous chapter.

THE PRINCIPLES OF SEMI-EMPIRICAL MOLECULAR-ORBITAL METHODS[b, 107]

The coordination chemist is largely concerned with complexes in which the central metal ion is surrounded by a number of ligands, commonly four, five or six, which may be the same or different. Symmetry plays an important role in the application of semi-empirical methods to such complexes as in the crystal-field approach. In order to separate the role of symmetry from the essential character of the MO method, it is convenient to begin our discussion with a metal ion bonded to a single ligand. We are concerned in this section, therefore, with a hypothetical complex ML.

We saw in the early part of chapter 6 that the Hamiltonian for a many-electron system could be replaced by a suitable one-electron function. In this guise the various Coulomb and exchange interactions appear as an effective potential so that the molecular orbitals describing the behaviour of the electrons are solutions to the equation:

$$\mathcal{H}\psi_i = \epsilon\psi_i. \tag{8.3}$$

Unlike the Shulman–Sugano[97] or Watson–Freeman[116] calculations, the precise form of the Hamiltonian is not required: indeed, in the simpler Hückel MO theory $\mathcal{H}$ is left completely undefined.[107]. However, if we assume that an electron in the vicinity of one of the two nuclei is described essentially by an orbital of that atom, then we are implicitly assuming that the molecular Hamiltonian can be represented approximately by a superposition of the Hamiltonians of the two atoms. Such a situation exists in the form of the ionic Hamiltonian given by (6.18). This is also the physical basis of the linear combination of atomic orbitals (LCAO) approximation. While this approximation is conceptually useful, particularly in discussions of chemical bonds, it is by no means as necessary to the molecular-orbital formalism as some texts imply. We could, using the expansion theorem discussed earlier, construct the molecular orbitals in terms of the complete set of atomic orbitals centred on either nucleus. The main practical problem is that a very large basis set of atomic orbitals would be required to represent the electron behaviour near the other nucleus. The LCAO approximation enables us to represent all of the molecular orbitals by a relatively small number of suitably chosen atomic functions and so it is more convenient.

In the LCAO approximation, we construct MOs for the ML molecule as:

$$\psi_i = a\phi_{\mathrm{M}} + b\phi_{\mathrm{L}}, \tag{8.4}$$

where the ψ_i are solutions to (8.3) and ϕ_{M} and ϕ_{L} are atomic orbitals of the metal and ligand. The energies E of the molecular orbitals and the values of the coefficients a and b are obtained by solution of the secular determinant whose order is given by the number of atomic orbitals taken. For illustration, restriction to a single orbital from each atom yields the 2×2 determinant:

$$\begin{vmatrix} H_{\mathrm{M}}-E & H_{\mathrm{ML}}-SE \\ H_{\mathrm{LM}}-SE & H_{\mathrm{L}}-E \end{vmatrix} = 0, \tag{8.5}$$

where

$$H_{\mathrm{M}} = \int \phi_{\mathrm{M}}^{*} \mathscr{H} \phi_{\mathrm{M}} d\tau, \text{ energy of } \phi_{\mathrm{M}} \text{ in molecular environment;}$$

$$H_{\mathrm{L}} = \int \phi_{\mathrm{L}}^{*} \mathscr{H} \phi_{\mathrm{L}} d\tau, \text{ energy of } \phi_{\mathrm{L}} \text{ in molecular environment;}$$

$$H_{\mathrm{LM}} = H_{\mathrm{ML}} = \int \phi_{\mathrm{M}}^{*} \mathscr{H} \phi_{\mathrm{L}} d\tau, \text{ resonance integral;}$$

$$S = \int \phi_{\mathrm{M}}^{*} \cdot \phi_{\mathrm{L}} d\tau, \text{ overlap integral.}$$

The Hamiltonian $\mathscr{H}$ contains the kinetic energy operator and the average effective potential due to both nuclei and other electrons. We also assume ϕ_{M} and ϕ_{L} to be separately normalized and that their symmetry representations are such that S is non-zero. Apart from the linear combination of metal and ligand orbitals, the only assumption made in writing down such a determinant is that a variational procedure is used to give the coefficients when the energies are minimized. Further approximations which may be used to solve the determinant give rise to a family of semi-empirical molecular-orbital methods.

APPROXIMATION 1: THE HÜCKEL MODEL[107]

The simplest approximation we can use is based on the Hückel MO theory as applied to heteroatomic molecules. Here, the energies H_{M} and H_{L} are *parameterized* as α_{M} and α_{L}, and the exchange integral as β_{ML}. In the simplest form of Hückel theory the overlap integral is set to zero. This clearly drastic assumption can be avoided by using orthogonalized atomic orbitals for ϕ_{M} and ϕ_{L}, but then α and β take on slightly different meanings. The neglect of S is not necessary, however, and is not assumed in extended Hückel models. If it is neglected, to reduce the degree of parameterization, (8.5) becomes:

$$\begin{vmatrix} \alpha_{\mathrm{M}} - E & \beta_{\mathrm{ML}} \\ \beta_{\mathrm{ML}} & \alpha_{\mathrm{L}} - E \end{vmatrix} = 0, \tag{8.6}$$

and α and β are commonly referred to as Coulomb and resonance (or exchange) integrals. In organic systems, these parameters are approximately transferable from one molecule to another as the carbon atoms are roughly neutral in all cases and the $2p\pi$ overlaps lead to essentially constant resonance integrals. The theory then has some utility in correlating the relative molecular-orbital energies in terms of β. In our heteroatomic complex ML, and to a lesser extent in heteroorganic compounds, the energies of the atomic orbitals range widely, depending on relative atomic charges. Atomic orbital energies are thus not transferable from complex to complex even when the metal remains the same and ligands are somewhat similar. Since there is no way of relating the resonance integrals to variations in Coulomb integrals, the application of the Hückel model to metal complexes provides no information about the strength of metal–ligand interactions.

APPROXIMATION 2: THE WOLFSBERG–HELMHOLZ MODEL[117, a, b, 6]

The solution of the determinant (8.5) under the conditions that the overlap integral be non-zero and that H_M and H_L are in general widely different was accomplished by Wolfsberg and Helmholz[117] in their classic paper on MnO_4^-, CrO_4^{2-} and ClO_4^-. In the WH model the H_M and H_L are set equal to the ionization energies of the electrons described by ϕ_M and ϕ_L in the field set up by the nuclei and remaining electrons. Since these ionization energies will differ from the free-ion or free-atom values they are adjusted so as to correspond to the energies for atoms of charges equal to those assigned to the atoms in the molecule. The diagonal matrix elements H_M and H_L are thus set equal to the Valence State Ionization Energies (VSIE) or Valence Orbital Ionization Potentials (VOIP). Each VSIE is a function of the charge and orbital configuration of the atom in question. For a given configuration a plot of ionization energies for valence states of integral atomic charges is obtained from atomic spectral data and the values for fractional or non-integral charges are obtained by interpolation. The practical procedure has usually involved an iterative technique in which the fractional charges upon which the VSIEs depend are obtained by diagonalization of the secular matrix they determine.

The off-diagonal matrix element H_{ML} is assumed to be proportional to both the overlap between the two orbitals ϕ_M and ϕ_L and their relative energies. This recipe for the resonance integrals follows from an earlier approximation by Mulliken[72] concerning the partition of 'overlap charge density' in MO calculations. The approximation, which we do not discuss further here, is that:

$$\begin{aligned}\langle\phi_A|\mathscr{H}|\phi_B\rangle &\sim \tfrac{1}{2}S[\langle\phi_A|\mathscr{H}|\phi_A\rangle+\langle\phi_B|\mathscr{H}|\phi_B\rangle]\\ &= \tfrac{1}{2}S(H_A+H_B).\end{aligned} \tag{8.7}$$

The WH model uses a proportionality factor F such that H_{ML} is given by:

$$H_{ML} = F.S_{ML}\left(\frac{H_M+H_L}{2}\right), \tag{8.8}$$

where F is approximately 2.00. Equation (8.8) is often referred to as the Wolfsberg–Helmholz approximation even though the WH model involves more than this single assumption.

We do not discuss the success or failure of the WH model: a critical appraisal is given in reference (*a*). We are principally concerned with this model as providing a basis for the angular overlap model which further assumes that the covalent interaction is weak. Those interested in details of the application of the WH model will find the book by Ballhausen and Gray[b] a useful guide: an extremely lucid account is also given by Bedon, Horner and Tyree[6] who apply the model to TiF_6^{3-}. In summary, the application of the WH model requires knowledge of the ionization energies and the overlap integrals of the atomic orbitals involved in bonding. The diagonal matrix elements are obtained from experimental data rather than by calculation, so that the solution of the determinant (8.5) essentially reduces to the evaluation of overlap integrals.

APPROXIMATION 3: THE ANGULAR OVERLAP MODEL[94, 95, c]

Because of the highly heteropolar bonding involved in transition metal coordination (at least, compared with the carbon chemistry area), that is H_M and H_L are very different, it may be assumed that the covalent bonding interaction is weak. Such terms are relative, of course, but the assumption has been used to permit simplification of the secular determinant by a perturbation theory argument.

If H_M and H_L are very different, the two roots of the determinant (8.5) are close to H_M and H_L. Assuming the ionization potential of the metal electron is less than that of the ligand electron, as will generally be the case, we have $E_1 \sim H_M$ and we put $E = H_M$ in all terms of the determinant except $H_M - E$: (8.5) then becomes:

$$\begin{vmatrix} H_M - E_1 & H_{ML} - S_{ML} H_M \\ H_{ML} - S_{ML} H_M & -(H_M - H_L) \end{vmatrix} = 0, \tag{8.9}$$

[note that $H_L - E$ becomes $-(H_M - H_L)$]. Then:

$$E_1 = H_M + \frac{H_{ML}^2 - 2H_{ML} H_M S_{ML} + S_{ML}^2 H_M^2}{H_M - H_L}. \tag{8.10}$$

Similarly, the second root $E_2 \sim H_L$ is given by:

$$E_2 = H_L - \frac{H_{ML}^2 - 2H_{ML} H_L S_{ML} + S_{ML}^2 H_M^2}{H_M - H_L}. \tag{8.11}$$

Equations (8.10) and (8.11) may be written in a more convenient form by making use of the Wolfsberg–Helmholz approximation for H_{ML}, to give:

$$E_1 = H_{\mathrm{M}} + \frac{[(\frac{1}{2}F-1)H_{\mathrm{M}} + \frac{1}{2}FH_{\mathrm{L}}]^2 S_{\mathrm{ML}}^2}{H_{\mathrm{M}} - H_{\mathrm{L}}}, \tag{8.12}$$

and

$$E_2 = H_{\mathrm{L}} - \frac{[(\frac{1}{2}F-1)H_{\mathrm{L}} + \frac{1}{2}FH_{\mathrm{M}}]^2 S_{\mathrm{ML}}^2}{H_{\mathrm{M}} - H_{\mathrm{L}}}, \tag{8.13}$$

A further simplification is possible if F is set exactly to 2.00, for then,

$$E_1 = H_{\mathrm{M}} + \left(\frac{H_{\mathrm{L}}^2}{H_{\mathrm{M}} - H_{\mathrm{L}}}\right) S_{\mathrm{ML}}^2, \tag{8.14}$$

and

$$E_2 = H_{\mathrm{L}} - \left(\frac{H_{\mathrm{M}}^2}{H_{\mathrm{M}} - H_{\mathrm{L}}}\right) S_{\mathrm{ML}}^2. \tag{8.15}$$

Now E_1 and E_2 are the energies of the molecular orbitals in the diatomic molecule ML, the antibonding and bonding orbitals respectively. Thus, *the energy by which the metal orbital is raised as a result of covalent bonding is directly proportional to the square of the diatomic overlap integral.* This conclusion forms a basic assumption of the angular overlap model.

It seems worth emphasizing at this point the role of overlap integrals in semi-empirical MO methods. While overlap and atomic orbital energies will both affect MO energies in the WH model and their effects cannot be completely independent, we can explore their separate effects by the following general discussion. Expanding the determinantal equation (8.5) gives the quadratic equation:

$$E^2(1 - S_{\mathrm{ML}}^2) + E(2S_{\mathrm{ML}}H_{\mathrm{ML}} - H_{\mathrm{M}} - H_{\mathrm{L}}) + (H_{\mathrm{M}}H_{\mathrm{L}} - H_{\mathrm{ML}}^2) = 0, \tag{8.16}$$

so that

$$E_1 + E_2 = \frac{H_{\mathrm{M}} + H_{\mathrm{L}} - 2S_{\mathrm{ML}}H_{\mathrm{ML}}}{1 - S_{\mathrm{ML}}^2}, \tag{8.17}$$

where E_1 and E_2 are the roots of (8.16). If the diagonal sum rule is obeyed (the sum of the eigenvalues of a secular determinant equals the sum of its diagonal elements) we have

$$E_1 + E_2 = H_{\mathrm{M}} + H_{\mathrm{L}}, \tag{8.18}$$

which is only valid when

$$\frac{H_{\mathrm{M}} + H_{\mathrm{L}} - 2S_{\mathrm{ML}}H_{\mathrm{ML}}}{1 - S_{\mathrm{ML}}^2} = \frac{H_{\mathrm{M}} + H_{\mathrm{L}}(1 - S_{\mathrm{ML}}^2)}{1 - S_{\mathrm{ML}}^2}, \tag{8.19}$$

or more simply, when

$$H_{ML} = \tfrac{1}{2}.S_{ML}(H_M + H_L). \tag{8.20}$$

However, if the H_{ML} are given according to the WH approximation then the diagonal sum rule is no longer obeyed if F is different from unity. [It is trivial that (8.18) follows from (8.17) if S_{ML} is zero.] Clearly, if F has its normal value of between 1.0 and 2.0 then the sum $E_1 + E_2$ is less negative† then $H_M + H_L$. It follows that, since the bonding molecular orbital lies lower than the lowest atomic orbital (of the ligand), the antibonding orbital has its energy raised by more than the bonding orbital has its energy lowered.

From (8.14) and (8.15) and re-emphasizing that H_M and H_L are very different with $H_L < H_M$, and that the overlap term is a small correction only, $S^2_{ML} \ll 1.0$ (overlap integrals between metal $3d$ orbitals and ligand $2s$, $2p$ or $3s$, $3p$ orbitals are often *ca.* 0.2–0.3) we see that the molecular-orbital energy levels are largely determined by the energies of the interacting atomic orbitals. The determination of the VSIEs is of great importance, therefore, especially because these energies are rapidly changing functions of the assumed charge distribution in the molecule. These charge distributions also determine the radial forms of the wavefunctions as expressed by the orbital exponents and hence also the overlap integrals. While this influence may be of secondary importance compared to the determination of H_M and H_L, it can be seen from (8.14) and (8.15) to produce a larger effect in E_1 than in E_2 as $H^2_L \gg H^2_M$. In other words, the sum $E_1 + E_2$ is greater than $H_M + H_L$ by an amount dependent on the overlap integral and, as E_1 is more susceptible, small changes in the overlap integrals will mainly change the energies of the antibonding orbitals. As crystal-field splittings in a molecular-orbital framework are associated with differences between antibonding orbitals, we see the origin of the emphasis on overlap integrals in the present MO calculations.

ANGULAR OVERLAP INTEGRALS

The angular overlap model takes its name and gains its power from the factorization of the overlap integrals in (8.14) and (8.15) into radial and angular parts. We write the overlap integrals as:

$$S_{ML} = F^l_\lambda . S^*_{ML}, \tag{8.21}$$

where the radial part S^*_{ML} is a function of the radial properties of metal and ligand orbitals and of bond length only. The angular part F^l_λ is a

† Assuming energies defined down from ionization limit at $E = 0$.

property of the orientations of the overlapping orbitals. It is called the angular overlap integral and is associated with the overlap of a metal orbital of azimuthal quantum number l with a ligand orbital having λ symmetry† with respect to the metal–ligand bond.‡ Since the antibonding energy E^* ($= E_1 - H_M$) is proportional to the square of the overlap integral, we have:

$$E^* = e_\lambda (F_\lambda^l)^2, \tag{8.22}$$

where

$$e_\lambda = \left(\frac{H_L^2}{H_M - H_L}\right) S_{ML}^{*2} \tag{8.23}$$

represents a geometry independent function (excluding bond length, of course). As all the metal 3d orbitals have the same value of H_M, the value of e_λ for a given ligand and metal–ligand distance is constant so that *the various antibonding effects for the d orbitals are proportional to the angular overlap integrals squared.*

We now discuss how explicit forms for the angular overlap integrals may be derived. This is simply a matter of geometry. The radial function written as S_{ML}^* expresses a geometry-independent overlap and is a pure diatomic overlap integral; for example, $S(p\sigma, \sigma)$, $S(d\sigma, \sigma)$ or $S(d\pi, \pi)$. The overlap integral between two generally oriented orbitals is expressed as a function of the pure diatomic integrals by resolving the metal functions in the coordinate frame of the ligand orbitals (or, of course, *vice versa*). Figure 8.2 depicts the coordinate frames for metal (x, y, z) and ligand (x', y', z'). The polar coordinates of the ligand in the metal frame are represented as (r, θ, ϕ) in which r is set as unity without loss of generality. The transformation we require is generally performed using the Eulerian rotation operators $R_z(\phi)$,

† The Greek letters universally used to represent types of bonding ($\sigma, \pi, \delta, \ldots$) refer to the component of the angular momentum of the MO about the internuclear axis. Thus, the orbitals are labelled by their value of λ, analogous to the quantum number m_l, as:

$$\begin{array}{l} \lambda = 0,\ 1,\ 2,\ 3,\ 4,\ \ldots \\ \quad\ \ \sigma,\ \pi,\ \delta,\ \phi,\ \gamma,\ \ldots. \end{array}$$

‡ This treatment of the angular overlap model is a simplified version of that due to Schäffer and Jørgensen,[95] and Schäffer.[94] Other variations are to be found in the literature. For example, the angular functions may be represented by a quantity Ξ_λ (Capital Greek xi) as in the original paper by Jørgensen and Schäffer. Ξ_λ and F_λ^l are related according to the equation

$$\Xi_\lambda = N_\lambda F_\lambda^l,$$

where N_λ takes the values:

$$\begin{aligned} \lambda = 0&: \quad (2l+1)^{\frac{1}{2}}, \\ \lambda = 1&: \quad [2(2l+1)l(l+1)]^{\frac{1}{2}}, \\ \lambda = 2&: \quad [\tfrac{1}{2}(2l+1)(l-1)l(l+1)(l+2)]^{\frac{1}{2}}. \end{aligned}$$

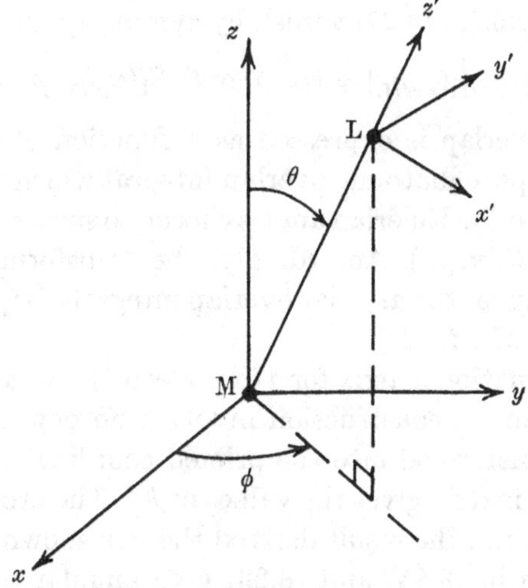

Figure 8.2. Axes for Eulerian rotations for an axially-symmetric M–L system. Note: x' lies in the plane defined by z and M–L.

$R_y(\theta)$, $R_z(\psi)$: the procedure is described in appendix 8A. The relative orientations of (x, y, z) and (x', y', z') are chosen so that the transformation matrix is as simple as possible. In the present case, no loss of generality (for an axially symmetric ligand) is suffered by rotation though ϕ, θ and 0 degrees, so that the transformation becomes:

$$\begin{pmatrix} x \\ y \\ z \end{pmatrix} = \begin{pmatrix} \cos\phi\cos\theta & -\sin\phi & \cos\phi\sin\theta \\ \sin\phi\cos\theta & \cos\phi & \sin\phi\sin\theta \\ -\sin\theta & 0 & \cos\theta \end{pmatrix} \begin{pmatrix} x' \\ y' \\ z' \end{pmatrix}. \tag{8.24}$$

We may now express a general diatomic overlap integral as an angular function of one or more pure overlap integrals. Consider a metal p_x orbital overlapping with a ligand $p_{x'}$ orbital. The total overlap integral S_{ML} is dependent upon the relative orientations of the orbitals (the angular overlap) and upon the characteristic n and l quantum numbers. Using (8.24), we have

$$x = x'\cos\phi\cos\theta - y'\sin\phi + z'\cos\phi\sin\theta, \tag{8.25}$$

so that

$$(p_x)_{\mathrm{M}} = (p_{x'})_{\mathrm{M}}\cos\phi\cos\theta - (p_{y'})_{\mathrm{M}}\sin\phi + (p_{z'})_{\mathrm{M}}\cos\phi\sin\theta. \tag{8.26}$$

The overlap integral we require, $S[(p_x)_{\mathrm{M}}(p_{x'})_{\mathrm{L}}]$, is thus given by:

$$\begin{aligned} S[(p_x)_{\mathrm{M}}(p_{x'})_{\mathrm{L}}] = {} & \cos\phi\cos\theta\,.\,S[(p_{x'})_{\mathrm{M}}(p_{x'})_{\mathrm{L}}] \\ & -\sin\phi\,.\,S[(p_{y'})_{\mathrm{M}}(p_{x'})_{\mathrm{L}}] \\ & +\cos\phi\sin\theta\,.\,S[(p_{z'})_{\mathrm{M}}(p_{x'})_{\mathrm{L}}]. \end{aligned} \tag{8.27}$$

The last two terms of (8.27) vanish by symmetry, leaving:

$$S[(p_x)_M(p_{x'})_L] = \cos\phi\cos\theta\,.\,S[(p_{x'})_M(p_{x'})_L], \tag{8.28}$$

in which the overlap is expressed as a function of the angular coordinate and a pure diatomic overlap integral which, in this case and corresponding to the Eulerian angles chosen to specify the transformation (8.24), is $S(p\pi, p\pi)$. Accordingly, the transformation matrix in (8.24) actually gives the angular overlap integrals for p orbitals and so gives the value of F^p_λ.

The transformation matrix for d (or indeed f) orbitals is rather more complicated, but its construction involves no new principles: the d orbitals are transformed into the primed coordinate system and the transformation matrix gives the values of F^d_λ. The process is described in appendix 8A and the result derived there is shown on p. 169.

The matrices in (8.25) and (8.29) give angular overlap integrals expressed in the simplest, *general* case for an axially symmetric ligand. Obviously, in the diatomic molecule ML we have been considering it is much simpler and more direct to use the *special* orientations of metal and ligand coordinate frames in which $z = z'$ and x and y are parallel to x' and y', respectively. Then we use the Eulerian rotations $\theta = 0$ and $\phi = 0$, so that (8.24) becomes:

$$\begin{pmatrix} x \\ y \\ z \end{pmatrix} = \begin{pmatrix} 1 & 0 & 0 \\ 0 & 1 & 0 \\ 0 & 0 & 1 \end{pmatrix} \begin{pmatrix} x' \\ y' \\ z' \end{pmatrix}. \tag{8.30}$$

The overlap integrals then have their maximum values, *viz*

$$S_{ML} = 1\,.\,S^*_{ML}. \tag{8.31}$$

We note that the absolute values of the F^l_λ are less than or equal to unity, of course.

Returning to the diatomic ML molecule, we finally write down the energies of the 'metal' orbitals relative to their non-bonding values, using (8.22) and the F^l_λ from (8.24) and (8.29) when $\theta = 0$ and $\phi = 0$:

$$\left.\begin{aligned} E^*(s) &= e_\sigma(s), \\ E^*(p_z) &= e_\sigma(p_z), \\ E^*(p_x) = E^*(p_y) &= e_\pi(p), \\ E^*(d_{z^2}) &= e_\sigma(d_{z^2}), \\ E^*(d_{xz}) = E^*(d_{yz}) &= e_\pi(d), \\ E^*(d_{x^2-y^2}) &= e_\delta(d). \end{aligned}\right\} \tag{8.32}$$

$$\begin{pmatrix} z^2 \\ yz \\ xz \\ xy \\ x^2-y^2 \end{pmatrix} = \begin{pmatrix} & A & \end{pmatrix} \begin{pmatrix} z'^2 \\ y'z' \\ x'z' \\ x'y' \\ x'^2-y'^2 \end{pmatrix}$$

where

$$A = \begin{pmatrix} \frac{1}{4}(1+3\cos 2\theta) & 0 & -\frac{1}{2}(\sqrt{3})\sin 2\theta & 0 & \frac{1}{4}(\sqrt{3})\,(1-\cos 2\theta) \\ \frac{1}{2}(\sqrt{3})\sin\phi\sin 2\theta & \cos\phi\cos\theta & \sin\phi\cos 2\theta & -\cos\phi\sin\theta & -\frac{1}{2}\sin\phi\sin 2\theta \\ \frac{1}{2}(\sqrt{3})\cos\phi\sin 2\theta & -\sin\phi\cos\theta & \cos\phi\cos 2\theta & \sin\phi\sin\theta & -\frac{1}{2}\cos\phi\sin 2\theta \\ \frac{1}{4}(\sqrt{3})\sin 2\phi(1-\cos 2\theta) & \cos 2\phi\sin\theta & \frac{1}{2}\sin 2\phi\sin 2\theta & \cos 2\phi\cos\theta & \frac{1}{4}\sin 2\phi(3+\cos 2\theta) \\ \frac{1}{4}(\sqrt{3})\cos 2\phi(1-\cos 2\theta) & -\sin 2\phi\sin\theta & \frac{1}{2}\cos 2\phi\sin 2\theta & -\sin 2\phi\cos\theta & \frac{1}{4}\cos 2\phi(3+\cos 2\theta) \end{pmatrix} \tag{8.29}$$

Equation (8.32) thus describes the antibonding energies in terms of radial *parameters* e_λ multiplied by numerical coefficients (in this, essentially trivial, case equal to unity) which are functions of the angular properties of the molecular coordination.

We have reviewed, without judgement, some of the various approximations which may be used to solve the 2×2 determinant in (8.5). The main principles have been illustrated by reference to a diatomic molecule ML. This was done in order to prevent the complications introduced by more highly symmetrical systems distracting attention from the essential principles of semi-empirical MO calculations. However, we now wish to move on to octahedral and tetrahedral molecules, and then distortions of these, in order to compare these models with crystal-field theory. In this book devoted to *parameters* we have hitherto minimized discussion of the essentially symmetry or group-theoretical parts of the subject. In crystal-field theory these are amply documented in standard texts, but this is probably less true for the symmetry side of semi-empirical MO methods in general and the angular overlap model in particular. Accordingly, we now devote some space to symmetry in MO theory but restrict ourselves largely to its use in the angular overlap model.

GROUP OVERLAP INTEGRALS

In previous sections we have been concerned with the overlap between orbitals of a single ligand and a metal ion. The diatomic overlap integral was defined as

$$S_{ij} = \int \phi_M \phi_L \, d\tau, \tag{8.33}$$

i and j referring to the metal and ligand. In a general complex ML_N the central metal orbitals overlap with some or all of the ligand orbitals in ways determined by symmetry. For the purpose of constructing molecular orbitals analogous to (8.4), ligand orbitals are grouped, or symmetry-adapted, in such a way that this group as a whole transforms as some irreducible representation of the point-group to which the molecular symmetry belongs: these symmetry-adapted ligand orbitals are then combined with a metal orbital transforming similarly. Thus, in (8.34)

$$\psi = a\phi_M + b\chi_L, \tag{8.34}$$

χ_L represents a general (normalized) combination of ligand atomic orbitals – a ligand group orbital. We may write,

$$\chi_L = \sum_{k=1}^{N} a_k \phi_k, \tag{8.35}$$

TABLE 8.1. *Ligand group orbitals for an octahedral complex*

Symmetry representation	Metal orbital	Ligand group orbital‡ σ-type†	π-type
a_{1g}	s	$\frac{1}{\sqrt{6}}(z_1+z_2+z_3+z_4+z_5+z_6)$	
e_g	$d_{x^2-y^2}$	$\frac{1}{2}(z_1-z_2+z_3-z_4)$	
	d_{z^2}	$\frac{1}{2\sqrt{3}}(2z_5+2z_6-z_1-z_2-z_3-z_4)$	
t_{1u}	p_x	$\frac{1}{\sqrt{2}}(z_1-z_3)$	$\frac{1}{2}(y_2+x_5-x_4-y_5)$
	p_y	$\frac{1}{\sqrt{2}}(z_2-z_4)$	$\frac{1}{2}(x_1+y_5-y_3-x_6)$
	p_z	$\frac{1}{\sqrt{2}}(z_5-z_6)$	$\frac{1}{2}(y_1+x_2-x_3-y_4)$
t_{2g}	d_{xz}		$\frac{1}{2}(y_1+x_5+x_3+y_6)$
	d_{yz}		$\frac{1}{2}(x_2+y_5+y_4+x_6)$
	d_{xy}		$\frac{1}{2}(x_1+y_2+y_3+x_4)$

† σ-type ligand orbitals are s or p_z.
‡ Normalized ignoring ligand–ligand overlap.

with $\sum_{k=1}^{N} a_k^2 = 1$. The procedures for obtaining symmetry adapted functions are well documented (see, for example, refs. 17, 2). In table 8.1 we list several such combinations for an octahedral molecule using the coordinate system shown in figure 8.3.

The general secular determinant is now written:

$$|H_{ij}-EG_{ij}| = 0, \tag{8.36}$$

where, analogous to (8.33), we define the group overlap integral

$$G_{ij} = \int \phi_i \chi_j d\tau. \tag{8.37}$$

The G_{ij} are expressed as multiples of the S_{ij}, the multiplying coefficients being simple numbers obtained by group theory and, as such, representing the molecular symmetry. Table 8.2 lists G_{ij} for octahedral and tetrahedral complexes in terms of the pure diatomic overlap integrals. The procedure by which this table was constructed is described in appendix 8 B.

As discussed above, the overlap integral furnishes a means of

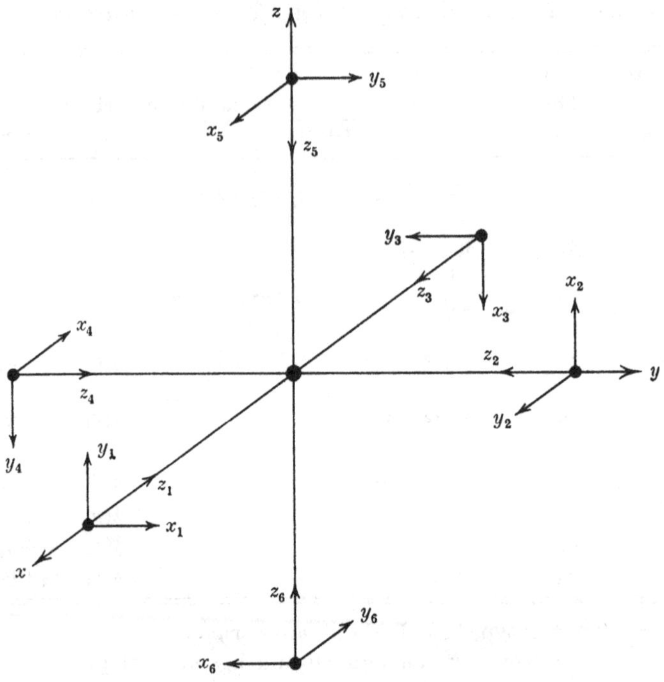

Figure 8.3. Coordinate system in O_h symmetry.

TABLE 8.2. *Group overlap integrals for octahedral and tetrahedral complexes*

Octahedral symmetry		Tetrahedral symmetry	
G_{ij}	$f(S_{ij})$	G_{ij}	$f(S_{ij})$
$G_{a_{1g}}(s,\sigma)$	$(\sqrt{6})S(s,\sigma)$	$G_{a_1}(s,\sigma)$	$2S(s,\sigma)$
$G_{e_g}(d,\sigma)$	$(\sqrt{3})S(d,\sigma)$	$G_e(d,\pi)$	$\frac{2\sqrt{2}}{\sqrt{3}}S(d,\pi)$
$G_{t_{2g}}(d,\pi)$	$2S(d,\pi)$	$G_{t_2}(p,\sigma)$	$\frac{2}{\sqrt{3}}S(p,\sigma)$
$G_{t_{1u}}(p,\sigma)$	$(\sqrt{2})S(p,\sigma)$	$G_{t_2}(p,\pi)$	$\frac{-2\sqrt{2}}{\sqrt{3}}S(p,\pi)$
$G_{t_{1u}}(p,\pi)$	$2S(p,\pi)$	$G_{t_2}(d,\sigma)$	$\frac{2}{\sqrt{3}}S(d,\sigma)$
		$G_{t_2}(d,\pi)$	$\frac{2\sqrt{2}}{3}S(d,\pi)$

assessing the extent of the interaction between the metal and ligand orbitals. In the WH model overlap effects are incorporated into the off-diagonal matrix elements: whereas Craig *et al.*[14] used overlap integrals as a direct measure of bond strength. In general semi-empirical MO methods, the relative energies of molecular orbitals are determined mainly by overlap integrals. In principle, as we are concerned with energy differences between antibonding orbitals, the '*d–d*' transitions between orbitals of the various symmetries can be rationalized by the magnitudes of the group overlap integrals, because the radial parameters $e_\lambda(d)$ are common factors. In particular, the splittings of the various '*d* orbitals' (for example, the $e_g^* - t_{2g}^*$ separation in the octahedron), which have the same atomic energies, are determined by the magnitudes of the group overlap integrals [in this case, $G_{e_g}(d, \sigma)$ and $G_{t_{2g}}(d, \pi)$] which express the differential overlap between the metal orbitals and the ligand orbitals. All that remains then are scaling factors – the radial overlap integrals – which put the energy splittings on an absolute scale. Anticipating later discussion, we can see already that the relative spectral splittings caused by low symmetry distortions will be dealt with by suitable group overlap integrals and scaling radial integrals; a comparison with the electrostatic model is therefore beginning to emerge.

THE ANGULAR OVERLAP MODEL IN ML_N COMPLEXES[96, 59, 94, 95]

In the next few sections we generalize our treatment of the linear ML system by incorporating the extra molecular symmetry often present in ML_N complexes *via* the group overlap integrals. We then illustrate the angular overlap model in detailed application to octahedral and tetrahedral molecules, and finally to some systems of lower symmetry.

We can generalize our discussion of the angular overlap model by noting that the orbitals of the central metal ion will overlap with ligand group orbitals rather than orbitals of a single ligand. The antibonding energies are now written as

$$E^* = \left(\frac{H_L^2}{H_M - H_L}\right) G_{ML}^2, \tag{8.38}$$

and so the antibonding effects are proportional to the squares of the group overlap integrals. As before, we separate angular and radial parts of the overlap integrals. In the cases where we use the symmetry-adapted ligand group orbitals the angular parts are simply the numerical coefficients relating the group overlap integrals to the pure diatomic overlap integrals. Then,

$$E^* = \left(\frac{H_{\mathrm{L}}^2}{H_{\mathrm{M}}-H_{\mathrm{L}}}\right)(cS_{\mathrm{ML}}^*)^2, \tag{8.39}$$

and, incorporating the S_{ML}^* into the first term once more, we have:

$$E^* = e_\lambda . c^2. \tag{8.40}$$

The c coefficients are taken from lists like table 8.2. Using that table we can immediately write down the σ- and π-antibonding effects in the octahedron, for example:

$$\left.\begin{aligned} E^*(e_g) &= 3e_\sigma, \\ E^*(t_{2g}) &= 4e_\pi. \end{aligned}\right\} \tag{8.41}$$

This method is especially convenient in those cases where the expression for the group overlap integrals (analogous to table 8.2) are easily evaluated. In the general case, however, the angular coefficients may be obtained simply by use of the F_λ^l matrices. In adopting this procedure it is assumed that the contributions from the ligands are additive, so that:

$$E^* = e_\lambda \sum_{k=1}^{N} (F_\lambda^l)^2, \tag{8.42}$$

assuming all N ligands to have the same radial parameters, i.e. that they are chemically identical and equidistant from the metal. The total antibonding effect on each metal orbital is then given as the sum of the contributions from each ligand. The total coefficient representing the angular overlap contributions from N ligands is found by summing the F_λ^l values obtained on inserting the ligand coordinates (θ_k, ϕ_k) one at a time into the F_λ^l matrices. We illustrate this method and show also the importance of ligand additivity in the following discussions.

THE OCTAHEDRON VERSUS THE TETRAHEDRON

The relative antibonding effects on the d orbitals of the metal and hence the orbital splittings are to be found by inserting the coordinates of the N ligands into the F_λ^l matrices and summing the various $(F_\lambda^l)^2$ values. For the octahedron, the six identical ligands situated at unit distance along the x, y, z axes have coordinates given in the table below. Consider the d_{z^2} orbital. From the F_λ^d matrix in (8.29) we have:

$$z^2 = [\tfrac{1}{4}(1+3\cos 2\theta)]z'^2 - [\tfrac{1}{2}(\surd 3)\sin 2\theta]x'z' + [\tfrac{1}{4}(\surd 3)(1-\cos 2\theta)] \quad (x'^2-y'^2). \tag{8.43}$$

Ligand	(x, y, z)	(θ, ϕ)
1	$(1, 0, 0)$	$(\frac{1}{2}\pi, 0)$
2	$(0, 1, 0)$	$(\frac{1}{2}\pi, \frac{1}{2}\pi)$
3	$(-1, 0, 0)$	$(\frac{1}{2}\pi, \pi)$
4	$(0, -1, 0)$	$(\frac{1}{2}\pi, \frac{3}{2}\pi)$
5	$(0, 0, 1)$	$(0, 0)$
6	$(0, 0, -1)$	$(\pi, 0)$

The orientations of the three equivalent orbitals described by the primed cartesian system are such as to lead to σ, π and δ bonding, respectively. The squares of the angular functions summed over the six ligands give the coefficients of e_σ, e_π and e_δ from the first, second and third terms. Thus we can write the total antibonding effect on the d_{z^2} orbital as:

$$\left\{\sum_{k=1}^{6} [\tfrac{1}{4}(1+3\cos 2\theta)]^2\right\} e_\sigma + \left\{\sum_{k=1}^{6} [-\tfrac{1}{2}(\sqrt{3})\sin 2\theta]^2\right\} e_\pi + \left\{\sum_{k=1}^{6} [\tfrac{1}{4}(\sqrt{3})(1-\cos 2\theta)]^2\right\} e_\delta. \quad (8.44)$$

Therefore, the energy $E^*(d_{z^2})$ is given (see table below) by:

$$E^*(d_{z^2}) = 3e_\sigma + 3e_\delta. \quad (8.45)$$

Ligand	$[\ldots]e_\sigma$	$[\ldots]e_\pi$	$[\ldots]e_\delta$
1	$[\frac{1}{4}(1+3\cos\pi)]^2$	$[-\frac{1}{2}(\sqrt{3})\sin\pi]^2$	$[\frac{1}{4}(\sqrt{3})(1-\cos\pi)]^2$
2	$[\frac{1}{4}(1+3\cos\pi)]^2$	$[-\frac{1}{2}(\sqrt{3})\sin\pi]^2$	$[\frac{1}{4}(\sqrt{3})(1-\cos\pi)]^2$
3	$[\frac{1}{4}(1+3\cos\pi)]^2$	$[-\frac{1}{2}(\sqrt{3})\sin\pi]^2$	$[\frac{1}{4}(\sqrt{3})(1-\cos\pi)]^2$
4	$[\frac{1}{4}(1+3\cos\pi)]^2$	$[-\frac{1}{2}(\sqrt{3})\sin\pi]^2$	$[\frac{1}{4}(\sqrt{3})(1-\cos\pi)]^2$
5	$[\frac{1}{4}(1+3\cos 0)]^2$	$[-\frac{1}{2}(\sqrt{3})\sin 0]^2$	$[\frac{1}{4}(\sqrt{3})(1-\cos 0)]^2$
6	$[\frac{1}{4}(1+3\cos 2\pi)]^2$	$[-\frac{1}{2}(\sqrt{3})\sin 2\pi]^2$	$[\frac{1}{4}(\sqrt{3})(1-\cos 2\pi)]^2$
$\Sigma =$	$[4\times\frac{1}{4}+1+1]$	$[0]$	$[4\times\frac{3}{4}+0+0]$

Repeating this procedure for the remaining d orbitals gives:

$$\left.\begin{aligned} E^*(d_{x^2-y^2}) &= E^*(d_{z^2}) = 3e_\sigma + 3e_\delta, \\ E^*(d_{xy}) &= E^*(d_{yz}) = E^*(d_{xz}) = 4e_\pi + 2e_\delta. \end{aligned}\right\} \quad (8.46)$$

Hence the orbital splitting

$$\Delta_{\text{oct}} = 3e_\sigma - 4e_\pi + e_\delta. \quad (8.47)$$

The four ligands in the tetrahedron have the coordinates:

ligand	(θ, ϕ)
1	$(\cos^{-1}\sqrt{\frac{1}{3}}, \frac{1}{4}\pi)$
2	$(\cos^{-1}\sqrt{\frac{1}{3}}, \frac{5}{4}\pi)$
3	$(\pi - \cos^{-1}\sqrt{\frac{1}{3}}, \frac{7}{4}\pi)$
4	$(\pi - \cos^{-1}\sqrt{\frac{1}{3}}, \frac{3}{4}\pi)$

The orbital energies may be shown to be:

$$\left.\begin{aligned} E^*(d_{z^2}) = E^*(d_{x^2-y^2}) &= \tfrac{8}{3}e_\pi + \tfrac{4}{3}e_\delta, \\ E^*(d_{xy}) = E^*(d_{xz}) = E^*(d_{yz}) &= \tfrac{4}{3}e_\sigma + \tfrac{8}{9}e_\pi + \tfrac{16}{9}e_\delta, \end{aligned}\right\} \tag{8.48}$$

giving
$$\Delta_{\text{tet}} = -\tfrac{4}{3}e_\sigma + \tfrac{16}{9}e_\pi - \tfrac{4}{9}e_\delta, \tag{8.49}$$

and the familiar 'geometrical' relationship:

$$\Delta_{\text{tet}} = -\tfrac{4}{9}\Delta_{\text{oct}}. \tag{8.50}$$

Note that the $-\frac{4}{9}$ ratio applies both to the overall Δ values and to the individual coefficients of the e_λ.

LOWER SYMMETRIES

For complexes of symmetry lower than cubic, a similar procedure to that above is used and the energies of the *d* orbitals similarly expressed as functions of the radial parameters. The linear diatomic molecule is such a case and we have previously listed the orbital energies in (8.32). However, for completeness we include the *d* orbital energies for this molecule in table 8.3 together with the results for other molecules which are of interest to us. We emphasize that for this present discussion we still assume that all ligands are identical and are at the same (unit) distance from the metal ion. In this table the orbitals are grouped according to their irreducible representations in each of the point groups considered; coordinates of the ligands are also indicated. It can be seen from this table that the sums of the e_σ columns equal N and the sums of the e_π and e_δ columns equal $2N$, where N is the number of ligands. Further, the sums of the e_λ contributions to each *d* orbital for each of the symmetry representations equal N. These results are examples of a general sum rule discussed by Schäffer and Jørgensen,[95,c] which we now discuss.

In the F_λ^l matrices each column represents the coefficients for different λ values of bonding. Thus in (8.29) the first column contains the F_σ^d values which are the angular overlap integrals or coefficients of the pure (d, σ) overlaps: they express the angular coefficients of the

TABLE 8.3. *The energies of the d orbitals in* ML_N *complexes*

d	$ML\,(C_{\infty v})$				$ML_2\,(D_{\infty h})$				$ML_3\,(D_{3h})$				$ML_4\,(D_{4h})$			
orbital		e_σ	e_π	e_δ		e_σ	e_π	e_δ		e_σ	e_π	e_δ		e_σ	e_π	e_δ
d_{z^2}	σ^+	1	0	0	σ_g^+	2	0	0	a_1'	$\frac{3}{4}$	0	$\frac{9}{4}$	a_{1g}	1	0	3
d_{yz}	π	0	1	0	π_g	0	2	0	e''	0	$\frac{3}{2}$	$\frac{3}{2}$	e_g	0	2	2
d_{xz}		0	1	0		0	2	0		0	$\frac{3}{2}$	$\frac{3}{2}$		0	2	2
d_{xy}	δ	0	0	1	δ_g	0	0	2	e'	$\frac{9}{8}$	$\frac{3}{2}$	$\frac{3}{8}$	b_{2g}	0	4	0
$d_{x^2-y^2}$		0	0	1		0	0	2		$\frac{9}{8}$	$\frac{3}{2}$	$\frac{3}{8}$	b_{1g}	3	0	1
ligand 1			$(0,0)$				$(0,0)$				$(\frac{1}{2}\pi, 0)$				$(\frac{1}{2}\pi, 0)$	
(θ,ϕ) 2							$(\pi,0)$				$(\frac{1}{2}\pi, \frac{2}{3}\pi)$				$(\frac{1}{2}\pi, \frac{1}{2}\pi)$	
3											$(\frac{1}{2}\pi, \frac{4}{3}\pi)$				$(\frac{1}{2}\pi, \pi)$	
4															$(\frac{1}{2}\pi, \frac{3}{2}\pi)$	

z'^2 components of the resolved atomic orbitals. Similarly in (8.24), the first column expresses the angular coefficients of the (p_x, π) bonding. In each case, the sum of the squares of the coefficients in each column of these unitary matrices is unity, representing the total angular bonding capacity of each of the λ orbitals of the ligand. Accordingly, the square of the coefficient of each metal orbital expresses the fraction of that orbital participating in the λ bond. Similarly, the sum of the squares of the coefficients in each row of the transformation matrices is unity, expressing the total angular bonding capacity of each of the $(2l+1)$ orbitals of the metal. If we combine this result with the additivity of ligand contributions, expressed by (8.42), we see that these summations remain equal to unity for each ligand taken in turn. In other words, for N ligands we have the rows and columns of the $(F_\lambda^l)^2$ matrices each summing to $1 \times N$. A general sum rule may now be defined. For each of the $(2l+1)$ metal orbitals, the sum of the coefficients of the e_λ equals the number of ligands. The specific results quoted in table 8.3 illustrate this rule, remembering that there are two sets each of ligand π and δ orbitals (i.e. two columns for the π and δ bonds) so that $\Sigma e_\pi = \Sigma e_\delta = 2N$.

Most powerfully, it follows from the additivity of ligand contributions that the energies of the d orbitals in an ML_N complex may be obtained as the sums of the energies in $ML_{N'}$ and $ML_{N''}$ where $N = N' + N''$. Thus addition of the bonding contributions from a ligand at (θ, ϕ) equal to $(0, 0)$ to those of the ligands in the ML_3 and ML_4 planar molecules produces the trigonal pyramid (C_{3v}) and the square pyramid (C_{4v}), respectively. The orbital energies for the molecules are given in table 8.4. Similarly, addition of the ML_2 orbital energies to those of ML_3 and ML_4 planes produce the trigonal bipyramid (D_{3h}) and the square bipyramid which, since all ligands are identical and at the same unit distance from the metal, is the octahedron. Table 8.5 lists the orbital energies for these two molecules.

TABLE 8.4. *Energies of d orbitals in the monopyramids C_{3v} and C_{4v}*

d orbital		ML_4 (C_{3v})				ML_5 (C_{4v})		
		e_σ	e_π	e_δ		e_σ	e_π	e_δ
d_{z^2}	a_1	$\frac{7}{4}$	0	$\frac{9}{4}$	a_1	2	0	3
d_{yz}	e	0	$\frac{5}{2}$	$\frac{3}{2}$	e	0	3	2
d_{xz}		0	$\frac{5}{2}$	$\frac{3}{2}$		0	3	2
d_{xy}	e	$\frac{9}{8}$	$\frac{3}{2}$	$\frac{11}{8}$	b_2	0	4	1
$d_{x^2-y^2}$		$\frac{9}{8}$	$\frac{3}{2}$	$\frac{11}{8}$	b_1	3	0	2

TABLE 8.5. *Energies of d orbitals in the bipyramids D_{3h} and O_h*

d orbital		ML_5 (D_{3h})				ML_6 (O_h)		
		e_σ	e_π	e_δ		e_σ	e_π	e_δ
d_{z^2}	a_1'	$\frac{11}{4}$	0	$\frac{9}{4}$	e_g	3	0	3
d_{yz}	e''	0	$\frac{7}{2}$	$\frac{3}{2}$		0	4	2
d_{xz}		0	$\frac{7}{2}$	$\frac{3}{2}$	t_{2g}	0	4	2
d_{xy}	e'	$\frac{9}{8}$	$\frac{3}{2}$	$\frac{19}{8}$		0	4	2
$d_{x^2-y^2}$		$\frac{9}{8}$	$\frac{3}{2}$	$\frac{19}{8}$	e_g	3	0	3

In certain symmetries it is possible that off-diagonal matrix elements appear which are not present in the component molecules; that is, a ligand λ orbital may interact with two non-degenerate d orbitals. An example is an off-diagonal element connecting the two e sets in $ML_4(C_{3v})$ in table 8.4, although in this case the matrix element vanishes. In general such matrix elements are evaluated using the general equation:

$$E^*(\phi_M \cdot \phi_{M'}) = \sum_{k=1}^{N} e_\lambda F_\lambda^l(\phi_M, \chi_k) \cdot F_\lambda^l(\phi_{M'} \cdot \chi_k). \tag{8.51}$$

DISTORTED MOLECULES

So far we have described ML_N complexes involving identical ligands and identical bond lengths and so all ligands have common radial parameters e_λ. If either of these two conditions is relaxed, e_λ parameters for one component will differ from those for the other. Such is the case in distorted molecules.

Let us consider molecules of the type $MA_4B_2(D_{4h})$ which we discussed in chapter 5. As we saw there, the quadrate symmetry is achieved by virtue of different M–A and M–B bond lengths or because ligands A and B are different, or, of course, by both causes. If A and B are identical but at different distances from the metal ion (e.g. the tetragonally-distorted octahedral $CuCl_6^{4-}$ ion) $e_\lambda(\mathrm{A})$ and $e_\lambda(\mathrm{B})$ will differ because the overlap integral S^*_{ML} is a function of bond length. On the other hand, if A and B are chemically different, H_{L} and S^*_{ML} will differ, again yielding different e_λ values. In either case, the orbital energies for the *trans*-MA_4B_2 molecule become:

$$\left.\begin{aligned} d_{z^2}(a_{1g})&:\quad e_\sigma(\mathrm{A})+2e_\sigma(\mathrm{B})+3e_\delta(\mathrm{A}),\\ \left.\begin{matrix} d_{yz}\\ d_{xz}\end{matrix}\right\}(e_g)&:\quad 2e_\pi(\mathrm{A})+2e_\pi(\mathrm{B})+2e_\delta(\mathrm{A}),\\ d_{xy}(b_{2g})&:\quad 4e_\pi(\mathrm{A})+2e_\delta(\mathrm{B}),\\ d_{x^2-y^2}(b_{1g})&:\quad 3e_\sigma(\mathrm{A})+e_\delta(\mathrm{A})+e_\delta(\mathrm{B}). \end{aligned}\right\} \tag{8.52}$$

In the angular overlap model, therefore, orbital splittings are functions of six parameters (term splittings also involve the interelectron repulsion parameter B) even in these highly symmetrical D_{4h} molecules. Frequently however, some of these parameter values may be set at zero. For example, when A is ammonia or A_2 is a diamine ligand, the nitrogen donor atoms are considered to have no orbitals available for π or δ bonding so that $e_\pi(\mathrm{A}) = e_\delta(\mathrm{A}) = 0$: if the axial B ligands are halides then $e_\delta(\mathrm{B}) = 0$. In these circumstances, orbital splittings involve three parameters only – $e_\sigma(\mathrm{A})$, $e_\sigma(\mathrm{B})$ and $e_\pi(\mathrm{B})$ – equivalent to the number of crystal field radial parameters required by symmetry (Dq, Ds, Dt say) and so their values may be obtained from the spectral transition energies. Other examples of bond length distortions have been treated in the same way. The $CuCl_5^{3-}$ ion is an example of a D_{3h} molecule with two axial bonds shorter than three equatorial ones: the two sets of chlorine ligands are assigned different e_σ and e_π parameters.

Angularly-distorted molecules are just as easily treated in the angular overlap model. If, as we assumed in chapter 4, bond lengths and ligands are identical and the radial parameters are unchanged by small angular distortions, the distortion is incorporated into the angular overlap integrals F^l_λ. Variations of the orbital energies as functions of distortion angle are simply calculated by varying that angle in the appropriate expressions for the F^l_λ coefficients. The molecules considered in tables 8.3, 8.4 and 8.5 are just special cases where the angular coefficients for the ligands are zero or simple multiples of $\frac{1}{2}\pi$ or $\frac{1}{3}\pi$. As in the electrostatic approach, angular distortion can be parameterized but this is not usually done in practice: instead atomic coordinates obtained by structure analyses are used in the evaluation of the F^l_λ matrices.

APPLICATIONS OF THE ANGULAR OVERLAP MODEL

Many of the early papers on the angular overlap model and its application were concerned with the splittings of the 4*f* orbitals in lanthanide complexes. These papers, which are discussed by Jørgensen,[c] generally considered the σ-antibonding effects only, so relating the *f* orbital energies to a single radial parameter. The *f* orbital splittings are then determined by the angular dependence of the overlaps. Quite complicated structures have been treated using this simpler model, including for example, cubic ML_8, octahedral and antiprismatic ML_6 and trigonal ML_9 systems. The angular overlap model is generally claimed to satisfactorily reproduce the experimental *f* orbital splittings using a single parameter.

Day and Jørgensen[19] have applied the angular overlap model to copper(II) chlorides and bromides of various structures: CuX_4^{2-} distorted tetrahedron D_{2d}, CuX_5^{3-} trigonal bipyramid D_{3h}, and CuX_4^{4-} tetragonally-elongated octahedron D_{4h}. Several divalent transition metal tetrahedral halides were also discussed. Again only σ-bonding effects were considered and so the energy levels of the copper(II) complexes were expressed in terms of $e_\sigma(S^*_{ML})^{-2}$. This form of parameterization was employed in order to take account of the different bond lengths in the various $CuCl_N$ and $CuBr_N$ complexes. Equal bond lengths were assumed in the tetrahedral and five-coordinate systems whereas different axial and equatorial bond lengths in the D_{4h} molecules were included in the angular parameters. However, this particular form of parameterization assesses the antibonding effects as being determined by the *energies* of the overlapping copper and halide orbitals. The use of a constant value for $e_\sigma(S^*_{ML})^{-2}$ (for the same halide atom) in all three complexes implies that H_M and H_L are not appreciably changed by small variations in bond lengths. Notwithstanding this (perhaps severe) approximation, the calculated orbital transition energies of the various copper(II) halide complexes are in moderately good agreement with the experimental values with $e_\sigma(S^*_{ML})^{-2} = 270\,000$ cm^{-1} for chloride and $e_\sigma(S^*_{ML})^{-2} = 226\,000$ cm^{-1} for bromide.

Schäffer[94(b)] has examined the effect of the small angular distortions which arise from chelation in molecules of the type tris(diamine) chromium(III) or cobalt(III) and analogous *cis* bis(diamine) complexes. The displacements of the donor atoms away from their ideal axial positions are described by two angular parameters (δ and ϵ). One of these (δ) describes the displacement of donor atoms away from two axes, say the donor atoms bonded along the positive x and y axes, in such a way as to change the angle of the chelate ring at the metal ion. The second parameter (ϵ) describes the tilt of the ligand away from that (xy) plane defined by two of the cartesian axes. This model was applied to the circular dichroism spectra of chelate complexes of the chromium(III) and cobalt(III) ions although this is obviously a way of discussing distortions due to chelation in general.

Hitherto, we have considered only those cases in which the ligand could be regarded as axially symmetric; for example, that π-bonding is the same in (local ligand) x- and y-directions. When this is not so, three Eulerian rotations are required and the F^l_λ matrices given in (8.24) and (8.29), for example, must be multiplied by matrices corresponding to a third rotation: the resultant matrix is referred to as a D^l_λ matrix by Schäffer.[94] Apart from this small algebraic complication, no new principles are involved.

More recently, Smith[98–102] has considered the d orbital splittings in non-cubic complexes of copper(II) as arising from the combination of σ- and π-antibonding effects and crystal-field electrostatic effects. In this combination of the angular overlap model and point-charge electrostatic theory, the d orbital energies in D_{4h} distorted octahedral molecules are given, in comparison with (8.52), by:

$$\left.\begin{aligned} d_{z^2}&: \; e_\sigma(\mathrm{A}) + 2e_\sigma(\mathrm{B}) + 6Dq - 2Ds - 6Dt, \\ d_{x^2-y^2}&: \; 3e_\sigma(\mathrm{A}) + 6Dq + 2Ds - Dt, \\ d_{xy}&: \; 4e_\pi(\mathrm{A}) - 4Dq + 2Ds - Dt, \\ d_{xz}, d_{yz}&: \; 2e_\pi(\mathrm{A}) + 2e_\pi(\mathrm{B}) - 4Dq - Ds + 4Dt, \end{aligned}\right\} \tag{8.53}$$

in which δ-bonding has been neglected. The procedure Smith uses to calculate the energies in (8.53) is essentially described by the following stages:

(*a*) The experimental transition energies are used to obtain a rough value for e_σ and, using (8.23), the value of H_{M} is found. For this purpose, H_{L} is taken as the ionization potential of the atomic orbital for a neutral ligand atom and a value of $(S^*_{\mathrm{ML}})^2$ is calculated using the neutral ligand orbital and the $3d$ radial wavefunction of Cu^+ (Richardson double-zeta function).

(*b*) The value of H_M is used to calculate n, the positive ionic charge on the copper atom, using the equation given by Jørgensen:

$$H_M = (20.1n^2 + 91.7n + 38.5) \times 10^3 \text{ cm}^{-1}. \quad (8.54)$$

This value of n is used to estimate $\langle r^2 \rangle$ and $\langle r^4 \rangle$ (the mean square and fourth power radii of the metal d orbitals) by interpolation of the $\langle r^n \rangle$ values calculated for Cu^0, Cu^+ and Cu^{2+}.

(*c*) The charge on the ligand atoms (ze) is calculated knowing the overall charge on the complex and the value of H_M. This charge, together with the $\langle r^n \rangle$ and the X-ray crystallographic bond lengths are then used to calculate the crystal-field radial parameters.

(*d*) The values of e_π are assumed to be related to those of e_σ by the relation:

$$\frac{e_\sigma}{e_\pi} = \frac{(S^*_{ML})^2_\sigma}{(S^*_{ML})^2_\pi}. \quad (8.55)$$

So that, by calculation of the appropriate overlap integrals, e_π is expressed as a function of e_σ. The relation (8.55) is valid insofar as the ligand atomic orbitals involved in σ and π bonding are of the same type (e.g.) both $2p$ functions say, rather than sp^2 for σ bonding and p for π bonding, for example.

The result of these assumptions is that the energies given in (8.53) can be expressed as functions of a single parameter e_σ. Smith further assumed that H_M and H_L could be taken as independent of the metal–ligand distance and so a single value of $e_\sigma(S^*_{ML})^{-2}$ (allowing for different bond lengths in the overlap integrals) was appropriate for a given ligand bonded to a given metal, irrespective of the symmetry of the complex. The model gave good fits between calculated and observed transition energies using a single parameter for each ligand type (i.e. one parameter for chlorides, another for bromides) regardless of the complex geometry. Thus a particular advantage was the transferability of radial parameters from one complex to another, at least within a restricted group of molecules. On the other hand several questions are raised by the procedure.

The first problem concerns the fact that, in (*a*) above, e_σ and H_M values were calculated from the observed transition energies. Even though rough values only are required at this stage, the procedure in (*a*) assumes a wholly covalent origin for the transition energies (i.e. *via* the angular overlap model) while Smith's final values ascribed nearly a quarter of the observed spectral splitting energies to the electrostatic terms. A similar problem concerns the fact that the charge distribution consistent with the H_M and H_L values used to estimate n differs from that used to calculate the diatomic overlap integrals and also from that used in the evaluation of the crystal-field radial parameters. However, as mentioned earlier, this inconsistency may be of little consequence if the overlap integrals are not much affected by variations in H_M and H_L; that is, in effect, that H_M and H_L are largely independent of metal–ligand distance. Clearly, simplifying assumptions are necessary in all semi-empirical models and those used here seem no more drastic than most. But, as discussed briefly in chapter 1, there are those who vociferously promote the angular overlap model relative to the electrostatic point-charge model on the grounds of 'reality'. We shall compare the two approaches shortly but with attention focussed on 'reality' it is worth noting a further point about Smith's approach. The electrostatic contribution was estimated from a relatively careful analysis of $\langle r^n \rangle$ values and charge distribution, the procedure having a very much more realistic look about it than the use of over-expanded Slater type orbitals implied in chapter 7. However, as pointed out in chapter 6, with inclusion of Kleiner's correction and of exchange terms, the electrostatic point-charge model must inevitably yield radial parameters of the wrong sign. This was ignored in chapter 7, for reasons carefully explained, in the spirit of empiricism. The same problem has been tacitly ignored by Smith. While not necessarily wishing to quarrel with his model, which has certainly shown much promise, we must note that the 'reality' in semi-empirical calculations seems to be a relative term.

Finally, Hitchman[53] has discussed the effect of a change in the metal ion while retaining identical ligands and molecular geometry. Among the examples considered were $M(acac)_2$, where M = Cu(II) or Ni(II) and acac = acetylacetone. The radial parameters e_σ and e_π differ from nickel to copper because of a lower Z_{eff} for nickel(II) relative to copper(II) so giving different values of H_M: also bond lengths are generally shorter in the nickel(II) complexes. With due allowance for these factors the angular overlap model was able to account for the different experimentally observed transition energies as arising from stronger metal–ligand interactions in the nickel(II) complexes.

PARAMETERS IN THE ANGULAR OVERLAP AND CRYSTAL-FIELD MODELS

We have already seen how the electrostatic crystal-field model and the angular overlap model both predict $\Delta_{tet} = -\frac{4}{9}\Delta_{oct}$. In that this relationship is not always exactly in agreement with experiment, both models are capable of explaining deviations from the rule due to changes in bond length, effective nuclear charge and ligand charges in fundamentally, if not apparently, similar ways. Our main interest in the present section is with the premises and philosophies of the two models rather than with their specific results. Traditional crystal-field theory assumes an electrostatic interaction between a metal ion and its surrounding ligands, represented by point-charges for example. The angular overlap model is based on weak covalent interactions between metal and ligands. These premises represent different viewpoints of the nature of the metal–ligand interaction. Both models, however, involve philosophies of empiricism and compromise and, in our view, are remarkably similar despite the obvious differences in premises and formalism.

Both models represent a first-order perturbation on the metal orbitals, the perturbation energies due to the ligand influence being caused either by bond formation or an electrostatic potential. The orbital energies are described by parameters which characterize the metal–ligand interaction and which are largely independent of the geometry of the complex. In the general case, for d orbitals say, the electrostatic model is parameterized by Cp and Dq values for each ligand (for a given metal) and the overlap model by e_σ, e_π and e_δ values, again different for each different ligand. The numerical coefficients for both sets of parameters are determined by angular or symmetry factors. Consider the breakdown of these parameters: for example, Dq and e_σ:

$$Dq = \tfrac{1}{6}(ze)e\left(\frac{\overline{r^4}}{a^5}\right) \quad \text{and} \quad e_\sigma = \frac{H_\mathrm{L}^2}{H_\mathrm{M}-H_\mathrm{L}}\cdot S_\mathrm{ML}^{*2}. \tag{8.56}$$

Dq and e_σ are both composite functions of the metal and ligand. The effective nuclear charge on the metal determines H_M on the one hand and $\overline{r^4}$ on the other: the ligand charge determines H_L and (ze), and the overall charge distribution determines S_ML^* and no doubt contributes also to the metal–ligand separation *via* Coulomb interactions. However, we emphasize again that these various properties are not independent of one another and attempts to assess their independent effects should be regarded as illustrative only. We stressed this in chapter 7, and also in the discussion near (8.16) in this chapter. For example, the variation of $10Dq$ with metal–ligand distance in NiO (determined from high-pressure spectroscopic data[105]) is consistent[103] with this parameter being proportional to $(S_\mathrm{ML}^*)^2$. The overlap integral was calculated for several metal–ligand distances while H_M and H_L were assumed constant. The same experimental data can also be interpreted[103] to show that Dq is inversely proportional to a^5 as required by crystal-field theory if r^n and a^{n+1} are considered as separable. Furthermore a large measure of agreement between the electrostatic[31] and overlap[102] models has been obtained for some eight-coordinate systems, for example.

In the angular overlap model, we might expect the radial parameters to have magnitudes in the order $e_\sigma > e_\pi > e_\delta$. Accordingly, several authors have considered σ-bonding effects only, even when the ligands have orbitals available for π-bonding. Alternatively we have seen how the assumption of initially equivalent σ- and π-orbitals gives the ratio of the radial parameters as the ratio of the squares of the appropriate overlap integrals. For copper(II)-chloride systems, the overlap integrals are 0.0768 for σ-bonding and 0.038 for π-bonding so that $e_\pi = 0.25 e_\sigma$. Empirical values of e_λ are certainly in the order required but it remains to be proved that their ratios are in general agreement with (8.55).

It is of interest to derive e_σ and e_π values for some MA_4B_2 molecules of D_{4h} symmetry and to explore relationships between them and Dq and Cp. The angular overlap parameters may be expressed in terms of McClure's parameters: following Lever[66] and neglecting contributions from δ-bonding, we have:

$$\left.\begin{aligned} d_\sigma &= -\tfrac{3}{4}[e_\sigma(b)-e_\sigma(a)], \\ d_\pi &= -[e_\pi(b)-e_\pi(a)], \end{aligned}\right\} \qquad (8.57)$$

where

$$\left.\begin{aligned} d_\sigma &= -\tfrac{12}{8}Ds-\tfrac{15}{8}Dt, \\ d_\pi &= -\tfrac{3}{2}Ds+\tfrac{5}{2}Dt, \end{aligned}\right\} \qquad (8.58)$$

giving

$$\left.\begin{aligned} Dq &= \tfrac{1}{10}(3e_\sigma-4e_\pi), \\ Cp &= \tfrac{2}{7}(e_\sigma+e_\pi). \end{aligned}\right\} \qquad (8.59)$$

These relationships hold for ligands with two, equivalent π-orbitals.

In table 8.6 we list values of e_σ and e_π deduced from the spectra of some nickel(II) complexes, together with Cp and Dq values derived therefrom using (8.59). Corresponding values are shown for the series $Cr(en)_2X_2$ obtained from experimental d_σ and d_π values, assuming no π-bonding from the equatorial, ethylenediamine ligands. The results of angular overlap calculations for the tetrahedral ions $CuCl_4^{2-}$ and $CuBr_4^{2-}$ are also given in the table together with the derived Cp/Dq ratios. Certain trends begin to emerge from these, admittedly limited, data and it seems appropriate here to try and 'predict' trends in the angular overlap parameters from simple chemical concepts of bonding.

TABLE 8.6. *Comparison of angular overlap and crystal-field parameters* (cm^{-1})

	Angular overlap parameters						Crystal-field parameters					
	Equatorial ligands (A)			Axial ligands (B)			Equatorial ligands			Axial ligands		
MA_4B_2 complex	e_σ	e_π	e_σ/e_π	e_σ	e_π	e_σ/e_π	Dq	Cp	Cp/Dq	Dq	Cp	Cp/Dq
[a]Ni pyridine$_4$ Cl_2	4670	570	8.19	2980	540	5.52	1173	1497	1.28	678	1006	1.48
Ni pyrazole$_4$ Cl_2	5480	1370	4.00	2540	380	6.68	1096	1957	1.79	610	834	1.37
Ni pyridine$_4$ Br_2	4500	500	9.00	2540	340	7.21	1150	1429	1.24	599	797	1.33
Ni pyrazole$_4$ Br_2	5440	1350	4.03	1980	240	8.25	1092	1940	1.78	498	634	1.27
[b]$Cr(en)_2F_2^+$	7233	—	—	8033	2000	4.02	2170	2067	0.95	1610	2867	1.78
$Cr(en)_2(H_2O)_2^{3+}$	7833	—	—	7497	1410	5.32	2350	2238	0.95	1685	2545	1.51
$Cr(en)_2Cl_2^+$	7500	—	—	5857	1040	5.63	2250	2143	0.95	1341	1971	1.47
$Cr(en)_2Br_2^+$	7500	—	—	5120	750	6.83	2250	2143	0.95	1236	1677	1.37

[a] M. A. Hitchman, *Inorg. Chem.* 1972, **11**, 2387. [b] L. Dubicki and P. Day, *Inorg. Chem.* 1971, **10**, 2043.

	Angular overlap[c]			Crystal-field		
Complex	e_σ	e_π	e_σ/e_π	Dq	Cp	Cp/Dq[d]
$CuCl_4^{2-}$	6764	1831	3.69	1297	2456	1.89
$CuBr_4^{2-}$	4616	821	5.62	1056	1553	1.47

[c] R. C. Slade, unpublished results.

[d] Cp/Dq values differ somewhat from those quoted in (4.21) which were derived allowing the distortion angle θ to vary parametrically.

From (8.12) we may write

$$e_\lambda = \left\{\frac{[(\frac{1}{2}F_\lambda - 1)H_M + \frac{1}{2}F_\lambda H_L]^2 S_\lambda^2}{H_M - H_L}\right\}, \tag{8.60}$$

where F_λ represents Wolfsberg–Helmholz proportionality constants for $\lambda(=\sigma,\pi,\delta,\ldots)$ bonding. Within Wolfsberg–Helmholz calculations, commonly used[a,b] values are $F_\sigma \sim 1.67$ and $F_\pi \sim 2.00$. With these values as guidelines we may explore some simple trends. In this we note that H_M and H_L are negative quantities and $|H_M| > |H_L|$. Other things being equal, we may deduce:

I *Decrease in* S_λ (e.g. bond length increase)
e_σ *and* e_π *will both decrease.* General consideration of overlap integrals as functions of bond lengths suggest that S_π falls off more rapidly with increasing bond length (at normal bond lengths) than S_σ. Hence e_π will decrease more quickly than e_σ and e_σ/e_π *increases*

II *Increase in* Z_{eff} (increase in $|H_M|$)
e_σ *and* e_π *will both increase* because the denominator in (8.60) decreases. For π-bonding the numerator is unchanged as $\frac{1}{2}F_\pi \sim 1$ but for σ-bonding $\frac{1}{2}F_\sigma \sim 0.835$ so that we have a contribution from H_M serving to decrease the numerator. Increasing $|H_M|$ thus decreases the numerator so that e_σ increases less quickly than e_π and so e_σ/e_π *decreases*

III *Decrease in ligand charge* (increase in $|H_L|$)
e_σ *and* e_π *will both decrease* as the denominator increases. Increasing $|H_L|$ will increase the numerator to a larger extent for π-overlap than for σ-overlap and hence e_σ/e_π *decreases.*

We give some simple examples of the application of trends I to III.

1. For $CuCl_4^{2-} \rightarrow CuBr_4^{2-}$:

I Overlap integrals decrease: e_λ decrease, e_σ/e_π increases
II Z_{eff} decreases: e_λ decrease, e_σ/e_π increases
III (ze) decreases: e_λ decrease, e_σ/e_π decreases.

Comparison with table 8.6 shows all three trends agree with experiment for e_λ values: we must assume trends I+II outweigh III so far as e_σ/e_π ratios are concerned.

2. For $Ni(py)_4Cl_2 \rightarrow Ni(pyrazole)_4Cl_2$.

Increasing bond length of axial Cl^- decreases e_σ and e_π and increases e_σ/e_π ratio.

3. For $Cr(en)_2X_2$.

Axial ligands decrease in e_σ and e_π values and increase in e_σ/e_π. This may be common behaviour for D_{4h} molecules.

Thus it appears that the observed trends in e_σ and e_π values are in line with simple chemical ideas of bonding, although it must be emphasized that the predicted trends above rely on many assumptions, not the least of which is that H_L values for σ- and π-bonding are equal. The discussion serves, however, to encourage confidence in the angular overlap parameters given in the table. It is clearly desirable to obtain further angular overlap parameters for a wide range of compounds and symmetries to put such confidence on a firmer basis and so help to establish any information about the chemical bonding these figures appear to suggest. One particular problem to be studied is the values and significance of e_σ and e_π values which relate to ligand orbitals not initially directed at the metal atom and subsequently resolved into σ- and π-components with respect to the M–L axis.

A particularly interesting point to emerge from the figures in table 8.6 is that Cp/Dq ratios in these molecules seem to be less than about 2, a conclusion which derives from both tetragonal octahedral and tetragonal tetrahedral systems. The evidence seems to be sufficiently persuasive to suggest re-investigation of those data in chapter 4 which were initially interpreted to favour higher Cp/Dq ratios. It was noted in chapter 4 that fits to the experimental data of the trigonally-distorted octahedral iron(II) and cobalt(II) molecules did allow low Cp/Dq ratios as well as high: questions about the ratio for the 'tetrahedral' $NiCl_4^{2-}$ system are then raised again (q.v.) and a relationship between Cp/Dq and coordination number is still to be established.

The angular overlap model may be said to have a more realistic physical basis in that it is concerned with the overlap of metal and ligand orbitals and we have seen in chapter 6 how the orbital splittings are largely determined by overlap effects. However, such a reasonable basis should not obscure the reliance of the theory on the often crude assumptions that are necessary if the radial parameters are to be quantitatively derived, when the assumptions inherent in the Wolfsberg–Helmholz model may be implicit. Finally we emphasise that the mathematical formulation of crystal-field theory, and in particular of equations like (4.12) defining the radial parameters Cp and Dq, does not of itself constitute a physical model for the calculation of these parameters. Indeed, Griffith[52] has pointed out that crystal-field theory should be regarded as an operator-equivalent formalism rather than as a genuine physical model. The qualitative treatment of ligand-field effects described in chapter 7, while mainly exploratory, does have some success in rationalizing the general behaviour of ligand-field radial integrals.

APPENDIX 8A. THE TRANSFORMATION PROPERTIES OF THE d-ORBITALS

In order to calculate the angular overlap integrals F^d_λ we need to transform or rotate the metal d orbitals defined in the cartesian frame (x, y, z) into equivalent orbitals (or combinations thereof) defined in the frame (x', y', z'). The primed frame is chosen so as to coincide with the local cartesian frame defined at the ligand apart from a translation along the metal–ligand axis which is taken as z'. We have already given the transformation matrix for the axes themselves and hence also the F^p_λ matrix, in (8.24). We now describe the construction of the F^d matrix given in (8.29).

A general method of transforming one cartesian system into another is by means of the Eulerian rotation operators $R_z(\phi)$, $R_y(\theta)$ and $R_z(\psi)$. The first operation is a rotation by an angle ϕ anticlockwise about the original z-axis; the second rotation is anticlockwise through θ about the new y-axis,† and the third rotation is anticlockwise through ψ about the new z axis. Thus rotating (x, y, z) by ϕ about z gives the new axes (x', y', z') as follows:

$$\begin{pmatrix} x' \\ y' \\ z' \end{pmatrix} = \begin{pmatrix} \cos\phi & \sin\phi & 0 \\ -\sin\phi & \cos\phi & 0 \\ 0 & 0 & 1 \end{pmatrix} \begin{pmatrix} x \\ y \\ z \end{pmatrix}. \tag{8A.1}$$

Alternatively the inverse transformation can be used if we wish to express (x, y, z) in terms of (x', y', z'):

$$\begin{pmatrix} x \\ y \\ z \end{pmatrix} = \begin{pmatrix} \cos\phi & -\sin\phi & 0 \\ \sin\phi & \cos\phi & 0 \\ 0 & 0 & 1 \end{pmatrix} \begin{pmatrix} x' \\ y' \\ z' \end{pmatrix}. \tag{8A.2}$$

Again following Schäffer[94(a)] we use the second alternative. Rotation about the new y'-axis gives the matrix:

$$\begin{pmatrix} x \\ y \\ z \end{pmatrix} = \begin{pmatrix} \cos\theta & 0 & \sin\theta \\ 0 & 1 & 0 \\ -\sin\theta & 0 & \cos\theta \end{pmatrix} \begin{pmatrix} x'' \\ y'' \\ z'' \end{pmatrix}. \tag{8A.3}$$

As $R_z(\psi)$ when $\psi = 0$ has no effect, the F^p_λ matrix given in (8.24) is given by the product of the transformations (8A.2) and (8A.3).

We now consider the analogous transformation properties of the d orbitals under $R_z(\phi)$ and $R_y(\theta)$. For this we require the detailed forms of the normalized cartesian d functions, as in table 8A.1.

† Ballhausen[3] gives the second rotation about the new x-axis. We here take this rotation about the new y-axis to retain consistency with Schäffer.[94(a)]

TABLE 8A.1. *The angular properties of the real cartesian orbitals*

Orbital	Abbreviation	Angular function†
$l = 0$	s	1
$l = 1$	p_x	$(\sqrt{3})x/r$
	p_y	$(\sqrt{3})y/r$
	p_z	$(\sqrt{3})z/r$
$l = 2$	d_{xz}	$(\sqrt{15})xz/r^2$
	d_{xy}	$(\sqrt{15})xy/r^2$
	d_{yz}	$(\sqrt{15})yz/r^2$
	$d_{x^2-y^2}$	$\frac{1}{2}(\sqrt{15})(x^2-y^2)/r^2$
	d_{z^2}	$(\sqrt{5})(z^2-\frac{1}{2}x^2-\frac{1}{2}y^2)/r^2$

† These functions are normalized to 4π. Some authors normalize to $4\pi/(2l+1)$ and some to unity.

Rotation under $R_z(\phi)$ is illustrated for the d_{xy} orbital. Using table 8A.1 and (8A.1):

$$\begin{aligned}R_z(\phi)|d_{xy}\rangle &= R_z(\phi)|(\sqrt{15})(xy)\rangle = (\sqrt{15})|x'y'\rangle \\ &= (\sqrt{15})(x\cos\phi + y\sin\phi)(-x\sin\phi + y\cos\phi) \\ &= (\sqrt{15})(-x^2\cos\phi\sin\phi + xy\cos^2\phi - xy\sin^2\phi \\ &\qquad + y^2\sin\phi\cos\phi) \\ &= (\sqrt{15})(xy)(\cos^2\phi - \sin^2\phi) - \tfrac{1}{2}(\sqrt{15})(x^2-y^2) \\ &\qquad (2\cos\phi\sin\phi) \\ &= \cos 2\phi \,.\, d_{xy} - \sin 2\phi \,.\, d_{x^2-y^2}. \end{aligned} \tag{8A.4}$$

Thus the d_{xy} and $d_{x^2-y^2}$ orbitals are mixed together as are the d_{yz} and d_{xz} orbitals: these results are shown in the $R_z(\phi)$ matrix, (8A.5), where the remaining elements are calculated in the same way:

$$\begin{pmatrix} z^2 \\ yz \\ xz \\ xy \\ x^2-y^2 \end{pmatrix} = \begin{pmatrix} 1 & 0 & 0 & 0 & 0 \\ 0 & \cos\phi & \sin\phi & 0 & 0 \\ 0 & -\sin\phi & \cos\phi & 0 & 0 \\ 0 & 0 & 0 & \cos 2\phi & \sin 2\phi \\ 0 & 0 & 0 & -\sin 2\phi & \cos 2\phi \end{pmatrix} \begin{pmatrix} z'^2 \\ y'z' \\ x'z' \\ x'y' \\ x'^2-y'^2 \end{pmatrix}. \tag{8A.5}$$

Rotation under $R_y(\theta)$ is illustrated by calculating the matrix element, $\langle x^2-y^2|R_y(\theta)|x^2-y^2\rangle$. We can cyclically permute the labels (x, y, z) and rewrite this element as $\langle y^2-z^2|R_z(\theta)|y^2-z^2\rangle$. The function $d_{y^2-z^2}$ is then expressed in terms of the usual d-orbitals. Using (8A.5) with ϕ put equal to θ, then allows evaluation of this matrix element. Thus:

$$(\surd 5)\,(x^2-\tfrac{1}{2}y^2-\tfrac{1}{2}z^2)/r^2 \equiv -\tfrac{1}{2}d_{z^2}+\tfrac{1}{2}(\surd 3)d_{x^2-y^2},$$

and

$$\tfrac{1}{2}(\surd 15)\,(y^2-z^2)/r^2 \equiv -\tfrac{1}{2}(\surd 3)\,d_{z^2}-\tfrac{1}{2}d_{x^2-y^2}, \qquad (8\text{A}.6)$$

so that $\langle y^2-z^2|R_z(\theta)|y^2-z^2\rangle$ becomes:

$$\langle -\tfrac{1}{2}(\surd 3)\,d_{z^2}-\tfrac{1}{2}d_{x^2-y^2}|R_z(\theta)|-\tfrac{1}{2}(\surd 3)\,d_{z^2}-\tfrac{1}{2}d_{x^2-y^2}\rangle$$
$$= \tfrac{3}{4}\langle d_{z^2}|R_z(\theta)|d_{z^2}\rangle+\tfrac{1}{4}(\surd 3)\,\langle d_{z^2}|R_z(\theta)|d_{x^2-y^2}\rangle$$
$$+\tfrac{1}{4}(\surd 3)\,\langle d_{x^2-y^2}|R_z(\theta)|d_{z^2}\rangle+\tfrac{1}{4}\langle d_{x^2-y^2}|R_z(\theta)|d_{x^2-y^2}\rangle. \qquad (8\text{A}.7)$$

The values of the four terms in (8A.7) are found from (8A.5) to be 1, 0, 0 and $\cos 2\theta$ so the required result is:

$$\langle d_{x^2-y^2}|R_y(\theta)|d_{x^2-y^2}\rangle = \tfrac{3}{4}+\tfrac{1}{4}\cos 2\theta. \qquad (8\text{A}.8)$$

The complete $R_y(\theta)$ matrix, obtained by similar means, is given in (8A.9).

$$\begin{pmatrix} z^2 \\ yz \\ xz \\ xy \\ x^2-y^2 \end{pmatrix} = \begin{pmatrix} \tfrac{1}{4}(1+3\cos 2\theta) & 0 & -\tfrac{1}{2}(\surd 3)\sin 2\theta & 0 & \tfrac{1}{4}(\surd 3)\,(1-\cos 2\theta) \\ 0 & \cos\theta & 0 & -\sin\theta & 0 \\ \tfrac{1}{2}(\surd 3)\sin 2\theta & 0 & \cos 2\theta & 0 & -\tfrac{1}{2}\sin 2\theta \\ 0 & \sin\theta & 0 & \cos\theta & 0 \\ \tfrac{1}{4}(\surd 3)\,(1-\cos 2\theta) & 0 & \tfrac{1}{2}\sin 2\theta & 0 & \tfrac{1}{4}(3+\cos 2\theta) \end{pmatrix} \begin{pmatrix} z'^2 \\ y'z' \\ x'z' \\ x'y' \\ x'^2-y'^2 \end{pmatrix} \qquad (8\text{A}.9)$$

The combined transformations (8A.5) and (8A.9) give the required transformation matrix $R_z(\phi).R_y(\theta)$ which is the F^d_λ matrix of (8.29).

APPENDIX 8B. EVALUATION OF GROUP OVERLAP INTEGRALS

The integrals expressing the overlap between orbitals of the central metal ion and the ligand group orbitals are labelled σ, π, δ etc. by considering the essentially cylindrical symmetry in the 'bonding regions'. Simple expression for these integrals as functions of the pure diatomic overlap integrals are obtained by resolving the metal orbitals into equivalent orbitals defined with respect to coordinates parallel to the local coordinates of each ligand: again, the metal–ligand bond direction is usually taken as the z'-axis. We illustrate the procedure with the tetrahedron and then show how simplification occurs in the case of the octahedron.

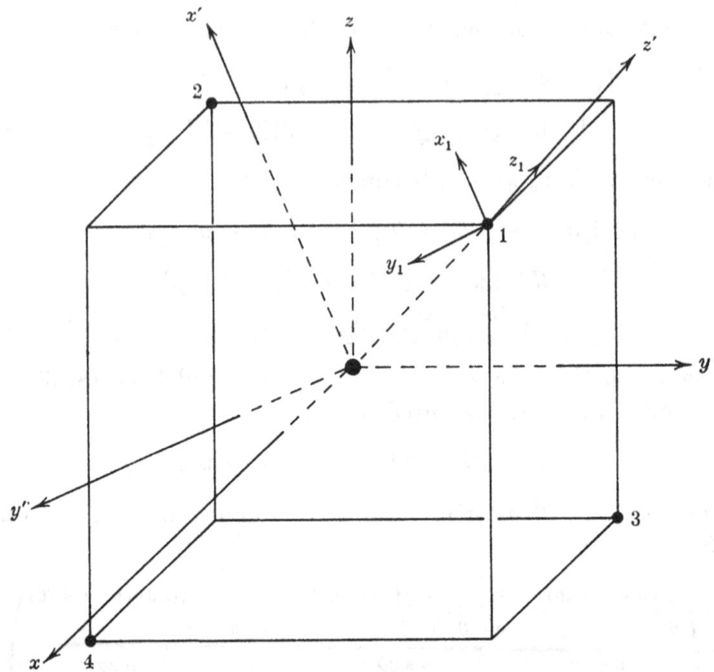

Figure 8B.1. Axes for Eulerian rotations in the tetrahedron.

THE TETRAHEDRON[b,3]

Ligand group orbitals, symmetry-adapted for bonding in a tetrahedral complex (analogous to those for the octahedron in table 8.1) are given by Wolfsberg and Helmholz,[117] and by Ballhausen and Gray:[b] we do not reproduce them here. However, one such ligand group orbital, composed of s orbitals, is $\frac{1}{2}(s_1+s_2-s_3-s_4)$ in the coordinate frame of figure 8B.1 and has t_2 symmetry. This group orbital overlaps with the metal d_{xy} orbital. The required group overlap integral $G_{t_2}(d, \sigma_s)$ is given as:

$$G_{t_2}(d, \sigma_s) = \int d_{xy} \cdot \tfrac{1}{2}(s_1+s_2-s_3-s_4)dv, \qquad (8B.1)$$

and since $-S(d_{xy}, s_3) = S(d_{xy}, s_1) = S(d_{xy}, s_2) = -S(d_{xy}, s_4)$ the total integral can be written as four times the overlap of d_{xy} with s_1. Thus,

$$G_{t_2} = 2S(d_{xy}, s_1). \qquad (8B.2)$$

We must now express the right-hand side of (8B.2) as a function of the pure diatomic overlap integral $S(d, \sigma)$. For this we express the d_{xy} orbital as the equivalent function in a new cartesian system (x', y', z') based on the local system at ligand 1 (x_1, y_1, z_1). This is shown in figure

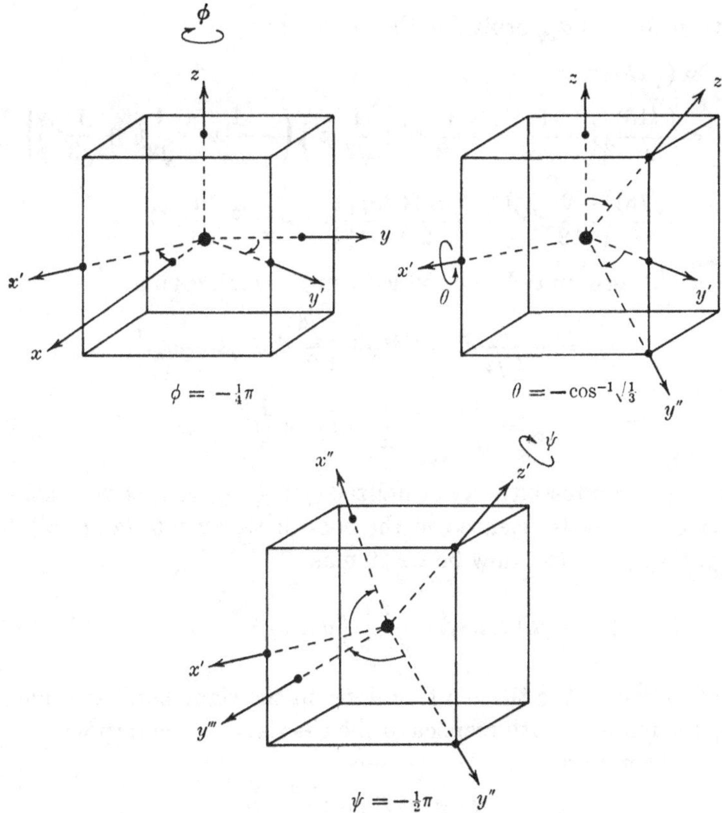

Figure 8B.2. Eulerian rotations in T_d symmetry.

8B.1. The two systems (x, y, z) and (x', y', z') are related† by the Eulerian angles

$$\phi = -\tfrac{1}{2}\pi, \quad \theta = -\cos^{-1}\sqrt{\tfrac{1}{3}} \quad \text{or} \quad -\sin^{-1}\sqrt{\tfrac{2}{3}}, \quad \psi = -\tfrac{1}{2}\pi \qquad (8\text{B}.3)$$

shown in figure 8B.2. Using these rotations we have the transformation:

$$\begin{pmatrix} x \\ y \\ z \end{pmatrix} = \begin{pmatrix} -\dfrac{1}{\sqrt{6}} & \dfrac{1}{\sqrt{2}} & \dfrac{1}{\sqrt{3}} \\ -\dfrac{1}{\sqrt{6}} & -\dfrac{1}{\sqrt{2}} & \dfrac{1}{\sqrt{3}} \\ \sqrt{\dfrac{2}{3}} & 0 & \dfrac{1}{\sqrt{3}} \end{pmatrix} \begin{pmatrix} x' \\ y' \\ z' \end{pmatrix}. \qquad (8\text{B}.4)$$

† In this appendix we take the second Eulerian rotation about the new x-axis. Also the (x, y, z) functions are written in terms of the (x', y', z') functions, again different from the conventions followed in appendix 8A. We do this in order to follow closely the cited authors in each case: we note that the procedures are equivalent and the final results do not depend on which conventions are used.

The transformed d_{xy} orbital is thus given by:

$$d_{xy} = (\surd 15)xy/r^2$$

$$= \frac{(\surd 15)}{r^2}\left\{\left(-\frac{1}{\surd 6}.x' + \frac{1}{\surd 2}.y' + \frac{1}{\surd 3}.z'\right)\left(-\frac{1}{\surd 6}x' - \frac{1}{\surd 2}y' + \frac{1}{\surd 3}z'\right)\right\}$$

$$= \frac{(\surd 15)}{r^2}\left\{\frac{-2}{\surd 18}x'z'\right\} + \frac{2(\surd 3)(\surd 5)}{2r^2}\left\{\frac{1}{6}x'^2 - \frac{1}{2}y'^2 + \frac{1}{3}z'^2\right\}.$$

The second term may be simplified using (8A.6), giving:

$$d_{xy} = -\frac{2}{3\surd 2}d_{x'z'} + 2(\surd 3)\left\{\frac{\surd 3}{9}d_{x'^2-y'^2} + \frac{1}{6}d_{z'^2}\right\},$$

i.e.
$$d_{xy} = \frac{1}{\surd 3}d_{z'^2} - \frac{2}{3\surd 2}d_{x'z'} + \frac{2}{3}d_{x'^2-y'^2}, \tag{8B.5}$$

so that d_{xy} is expressed as a normalized $(\frac{1}{3}+\frac{4}{18}+\frac{4}{9} = 1)$ combination of equivalent orbitals oriented in the coordinate system (x', y', z'). The integral $S(d_{xy}, s_1)$ can now be written as:

$$S(d_{xy}, s_1) = \frac{1}{\surd 3}S(d_{z'^2}, s_1) - \frac{2}{3\surd 2}S(d_{x'z'}, s_1) + \frac{2}{3}S(d_{x'^2-y'^2}, s_1). \tag{8B.6}$$

Only the first of the three d functions on the right-hand side has the correct symmetry with respect to the z'-axis for σ overlap with the s_1 orbital. Therefore,

$$S(d_{xy}, s_1) = \frac{1}{\surd 3}S(d, \sigma), \tag{8B.7}$$

giving
$$G_{t_2}(d, \sigma) = \frac{2}{\surd 3}S(d, \sigma), \tag{8B.8}$$

expressing the group overlap integral as an angular function $(2/\surd 3)$ of an appropriate pure diatomic overlap integral. The other expressions in table 8.2 are evaluated by similar means.

THE OCTAHEDRON

In the octahedron, the rotation of the coordinate system (x, y, z) into new systems appropriate to each ligand local system is facilitated because the ligands lie on the rotation axes of the Eulerian operators. The transformations may thus be written down directly by inspection. They are given in table 8B.1. The group overlap integrals for the metal e_g orbitals with the appropriate ligand group orbitals can be written:

$$G_{e_g}(d_{x^2-y^2}, \sigma) = \int (\surd 3)c.(x^2-y^2).\tfrac{1}{2}(\sigma_1 - \sigma_2 + \sigma_3 - \sigma_4)dv, \tag{8B.9}$$

TABLE 8B.1. *Coordinate transformations in O_h symmetry*

Transformed to ligand	x	y	z
1	z'	x'	y'
2	y'	$-z'$	x'
3	z'	$-y'$	$-x'$
4	$-x'$	z'	$-y'$
5	x'	y'	$-z'$
6	$-y'$	$-x'$	z'

where $c = (\sqrt{5})/2r^2$ and z_1, z_2, etc. from table 8.1 have been replaced by σ_1, σ_2 etc. to emphasize that no distinction is made between ligand $s\sigma$ and $p\sigma$ orbitals. Analogous to (8B.9) we can write:

$$G_{e_g}(d_{z^2}, \sigma) = \int c(2z^2 - x^2 - y^2) \cdot \frac{1}{2\sqrt{3}} \cdot (2\sigma_5 + 2\sigma_6 - \sigma_1 - \sigma_2 - \sigma_3 - \sigma_4)\, dv. \tag{8B.10}$$

Using table 8B.1, we transform the metal $d_{x^2-y^2}$ orbital to match each ligand in turn: since all ligands are alike we no longer require the subscripts. Thus:

$$\left.\begin{aligned} (x^2 - y^2)\sigma_1 &\to (z'^2 - x'^2)\sigma, \\ (x^2 - y^2)\sigma_2 &\to (y'^2 - z'^2)\sigma, \\ (x^2 - y^2)\sigma_3 &\to (z'^2 - y'^2)\sigma, \\ (x^2 - y^2)\sigma_4 &\to (x'^2 - z'^2)\sigma. \end{aligned}\right\} \tag{8B.11}$$

Summing over all four ligands then gives

$$\begin{aligned} G_{e_g}(d_{x^2-y^2}, \sigma) &= \int \tfrac{1}{2}(\sqrt{3})\, c(4z'^2 - 2x'^2 - 2y'^2) \cdot \sigma\, dv \\ &= (\sqrt{3}) \int c(2z'^2 - x'^2 - y'^2)\sigma\, dv. \end{aligned} \tag{8B.12}$$

Equation (8B.12) represents the σ overlap between a ligand orbital and a metal orbital of $d_{z'^2}$ type where z' is the metal–ligand axis. It therefore represents a pure $(d\sigma, \sigma)$ diatomic overlap. We therefore write (8B.12) as:

$$G_{e_g}(d_{x^2-y^2}, \sigma) = G_{e_g}(d, \sigma) = (\sqrt{3})\, S(d, \sigma). \tag{8B.13}$$

Similarly it may be shown that

$$G_{e_g}(d_{z^2}, \sigma) = (\sqrt{3})\, S(d, \sigma), \tag{8B.14}$$

giving the equivalent overlap with the ligand σ orbitals.

GENERAL TREATMENTS

The more general problem of overlaps between a metal orbital and the orbitals of N surrounding ligands in molecules of various symmetries has been considered by Dunitz and Orgel,[21] and more recently by Kettle.[61] General expressions are given in each case, relating the group overlap integrals to the pure diatomic integrals for the metal s, p and d orbitals as functions of the molecular geometry and the coefficients of the ligand group orbitals. These expressions are especially well suited to the evaluation of overlap integrals as functions of variable distortion angle.

DIATOMIC OVERLAP INTEGRALS

The final steps in an angular overlap calculation may involve the evaluation of the pure diatomic overlap integrals S^*_{ML}. In many cases these may be computed directly from available 'best' wavefunctions using an established computer program. However, for completeness' sake, we mention another source for these integrals. The classic paper of Mulliken *et al.*[73] lists extensive tables of diatomic overlap integrals expressed according to the parameters p and t:

$$p = \tfrac{1}{2}(\alpha_a + \alpha_b)R/a_0,$$
$$t = (\alpha_a - \alpha_b)/(\alpha_a + \alpha_b),$$

where α_a and α_b refer to exponents of Slater orbitals, R is the internuclear distance in Å and a_0 is the Bohr radius (0.529 Å). By convention, if the two Slater orbitals have different principal quantum numbers, that with the smaller value of n is labelled a. The tables of Mulliken *et al.* are used as follows. For the overlap of two Slater orbitals with exponents α_a and α_b on atoms distance R Å apart, values of p and t are calculated. The tables which list overlap integrals as functions of p and t only then give the required integral directly. Overlap integrals between orbitals represented as Watson SCF functions or Richardson double-zeta functions are expressed as a sum of the constituent Slater orbital overlaps weighted by the products of the appropriate expansion coefficients in metal and ligand functions.

GENERAL REFERENCES

(*a*) J. P. Dahl and C. J. Ballhausen, Molecular orbital theories of inorganic complexes, *Advances in Quantum Chemistry*, 1968, **4**, 170.

(*b*) C. J. Ballhausen and H. B. Gray, *Molecular Orbital Theory*, W. A. Benjamin Inc., New York, 1964.

(*c*) C. K. Jørgensen, *Modern Aspects of Ligand Field Theory*, North Holland, Amsterdam, 1971.

9
THE NEPHELAUXETIC EFFECT

As the principal concern of our introductory chapter was to distinguish the roles of symmetry on the one hand and splitting factors on the other, it was a matter of convenience to limit our considerations to Dq and the Spectrochemical Series. The other important electrostatic parameters in ligand-field theory, the interelectronic repulsion parameters, were thus temporarily ignored. While crystal-field integrals like Dq describe how the central metal electrons interact with the surrounding ligands, interelectron repulsion parameters are concerned with the metal electrons interacting with themselves. In any molecule or ion with more than one 'optical' electron or hole, both effects operate simultaneously, the metal electrons seeking to avoid each other and those of the ligands. The well-known correlation diagrams referring to 'weak-field' and 'strong-field' limits are concerned with the relative importance of these two effects. Near *both* extremes in these diagrams, degeneracies are removed to form levels called 'terms' and this identity of vocabulary suggests that these two effects appear on an equal footing. This is not so, of course, for the crystal-field operators are one-electron operators concerned with interactions between 'quantum' and 'classical' particles, while interelectron repulsion operators are of the two-electron type and are concerned with pairs of indistinguishable quantum particles. The detailed forms of these operators and of the interelectron repulsion parameters were discussed in chapter 3: for d electron level splittings we refer to the Condon–Shortley F_2, F_4 and the Racah B, C parameters defined there.

The Nephelauxetic effect describes the fact that the parameter values of interelectron repulsion are smaller in complexes than in the corresponding free-ions. As was the case for the Spectrochemical Series, we shall not present a complete phenomenological discussion of the Nephelauxetic effect but rather summarize some important features and theoretical proposals. Those interested in fuller details, especially of experimental data, might consult the many writings of Professor C. K. Jørgensen on this subject, several of which are cited at the end of this chapter.[a,b,c] Our discussion here is designed to explore the Nephelauxetic effect in the spirit adopted for our analysis of the crystal-field parameters in earlier chapters.

THE NEPHELAUXETIC SERIES[a,b,c]

At its simplest level, the Nephelauxetic effect may be described by reference to a representative interelectron repulsion parameter, values of which are deduced from the spectra of complexes in ways outlined later. That is, if B and C are assumed to be reduced from their free-ion values on complexation by similar proportions, the value of B in a complex may be taken to represent the Nephelauxetic effect in any given case. More usefully, we define β as the ratio of B in the complex to B_0 in the free-ion:

$$\beta = B/B_0. \tag{9.1}$$

Such Nephelauxetic ratios have been determined for a wide range of complexes with respect to variation of both metal and ligand, the results being collected into the Nephelauxetic Series. Like the Spectrochemical Series, the Nephelauxetic Series for ligands is roughly transferable from metal to metal; defined in order of increasing values of $(1-\beta)$ it is:

$$F^- < H_2O < \text{urea} < NH_3 < \text{en} < \text{ox}^{2-} < \underline{N}CS < Cl^- \lesssim CN^- < Br^- < I^- < \text{dtp}^-, \tag{9.2}$$

where ox = oxalate and dtp = $(C_2H_5O)_2PS_2^-$. The corresponding series for metals, less well defined is:

$$\text{Mn(II)} \sim \text{V(II)} < \text{Ni(II)} \sim \text{Co(II)} < \text{Mo(II)} < \text{Re(IV)} \sim \text{Cr(III)} < \text{Fe(III)} \sim \text{Os(IV)} < \text{Ir(III)} \sim \text{Rh(III)} < \text{Co(III)} < \text{Pt(IV)} \sim \text{Mn(IV)} < \text{Ir(VI)} < \text{Pt(VI)}. \tag{9.3}$$

Like the Spectrochemical series, the Nephelauxetic effect can be factorized into functions of ligand only and metal only:

$$(1-\beta) = h(\text{ligands}).k(\text{central ion}). \tag{9.4}$$

Table 9.1 lists some h and k values and compares them with the corresponding f and g values of the Spectrochemical Series.

Let us summarize the main trends in the Nephelauxetic Series.

(1) The ordering of ligands is quite different from that in the Spectrochemical Series and, indeed, to a considerable extent is in the opposite direction. In terms of the donor atoms there is a close parallel between the Nephelauxetic effect and Pauling's electronegativities:

$$\begin{array}{c} \text{decreasing electronegativity} \rightarrow \\ \text{F O N Cl C Br S I} \\ \text{increasing Nephelauxetic effect } (1-\beta) \rightarrow \end{array} \tag{9.5}$$

TABLE 9.1. *Factorization of the Nephelauxetic and Spectrochemical Series*

Ligands	f	h	Metal ions	$g(10^3\,cm^{-1})$	k
$6F^-$	0.9	0.8	V(II)	12.3	0.08
$6H_2O$	1.00	1.0	Cr(III)	17.4	0.21
6 urea	0.91	1.2	Mn(II)	8.0	0.07
$6NH_3$	1.25	1.4	Mn(IV)	23	0.5
3 en	1.28	1.5	Fe(III)	14.0	0.24
$3ox^{2-}$	0.98	1.5	Co(III)	19.0	0.35
$6Cl^-$	0.80	2.0	Ni(II)	8.9	0.12
$6CN^-$	1.7	2.0	Mo(III)	24	0.15
$6Br^-$	0.76	2.3	Rh(III)	27	0.30
$3dtp^-$	0.86	2.8	Re(IV)	35	0.2
			Ir(III)	32	0.3
			Pt(IV)	36	0.5

In general the Nephelauxetic series of ligands resembles a chemical concept of increasing covalency more than does the Spectrochemical Series. In particular, the order (9.2) corresponds well with the reducing power of the ligands; that is, their tendency to lose electrons:

$$F^- < H_2O < Cl^- < Br^- < I^- < O^{2-} < S^{2-}. \qquad (9.6)$$

(2) The correlation between reducing power and the Nephelauxetic effect for ligands in (9.6) is clear, with the exception of the oxide which is not normally considered to be more reducing than the iodide. This exception highlights the second general observation that the Nephelauxetic effect also depends on bond lengths. Oxides generally involve much shorter metal-oxygen bonds than aquo complexes. The greater Nephelauxetic effect associated with shorter bond lengths has been demonstrated elsewhere;[57] for example, for praseodymium–chlorine bonds in various lattices or from spectral studies of transition-metal and lanthanide complexes under high pressure.

(3) The Nephelauxetic effect for metals (9.3) resembles the analogous Spectrochemical Series in that increasing perturbations on the free-ions are associated with increasing oxidation state, but in other respects the series are quite different. Most particularly, while Dq values increase markedly along the series $3d^n < 4d^n < 5d^n$ for a given oxidation number, the Nephelauxetic effect $(1-\beta)$ shows little change and in fact a slight decrease. As far as the dependence of β on oxidation state is concerned we can discern, qualitatively at least, that the Nephelauxetic effect for metals increases as their oxidizing power.

(4) The roles of ligand reducing power and metal oxidizing power both describe transfer of electron density from ligands to metal. It is not surprising, therefore, to observe a close parallel between both ligand- and metal-Nephelauxetic series and the corresponding hyperchromic series describing increasing intensity of Laporte-forbidden ligand-field bands. Details may be found in references to Jørgensen.[a, b, c]

(5) Nephelauxetic effects in tetrahedral complexes have been reported as very similar, actually slightly larger, than in the corresponding octahedral complexes.

INTERPRETATIONS I

The earliest interpretations of the Nephelauxetic effect were due to Tanabe and Sugano, Owen, and Orgel and have since been elaborated by Jørgensen and others (see *c* and 15, 16). Two main mechanisms are described, referred to as 'central-field covalency' and 'symmetry-restricted covalency' effects.

Central-field covalency refers to transfer of negative charge from the ligands to the central metal ion, so tending to satisfy Pauling's electroneutrality principle and reflecting the reducing power of the ligands and oxidizing power of the metal, and reduces the effective nuclear charge on the metal. The resulting expansion of the metal orbitals, is supposed to increase the average distance between metal electrons and hence decrease interelectron repulsion parameters. While the central-field covalency effect is easily appreciated in qualitative terms, quantitative appraisals are still unclear. This is partly associated with the difficulties of separating central-field from symmetry-restricted covalency effects, but the question of bond lengths is clearly an important factor too. In addition Craig and Magnusson[15, 16] have discussed how the outer parts of metal wavefunctions may be differentially expanded with respect to the inner parts by the purely electrostatic effect of the surrounding ligands: we shall return to this matter later.

Symmetry-restricted covalency comments on the formation of molecular orbitals on complexation. Thus central-field covalency describes changes in Z_{eff} on the metal while symmetry-restricted covalency discusses electron delocalization effects. Reduction of the interelectron repulsion parameters, due to symmetry-restricted covalency effects, occurs in two main ways. One concerns the increased space available for the metal electrons, especially if we are considering them as housed in molecular orbitals formed with empty ligand acceptor orbitals. This idea presupposes that the interelectron repulsion parameters associated with the ligand are not larger than those on the metal or at least if they are, that the overall increase in space available for the electrons in the 'overlap regions' of a bond can overcome such an effect. Another mechanism stems from the fact that the orbital quantum number l is effectively reduced by admixture of ligand s or p functions into metal d or f functions. This takes place in the same way that gives rise to Stevens' orbital reduction factor k, used in the magnetic moment operator $\mu = \beta_0(kL+2S)$. A recent review[41] of orbital reduction factors also discussed how configuration mixing can reduce the effective l quantum number. For example, in tetrahedral symmetry the lack of a centre of inversion permits mixing of metal d and p functions which of itself will reduce k below unity although, as also described in the review, these effects are likely to be large only by a cooperative mechanism between covalency and configuration mixing.

Perhaps the most important aspect of symmetry-restricted covalency (certainly the most featured in the literature) is that it implies possible differential expansion among the central metal orbitals. Suppose we consider only octahedral or tetrahedral systems for the moment. A set of five d orbitals split into t_2 and e symmetry sets, the orbital-triplet being directed between the ligands in octahedral symmetry, for example, and the doublet directly at them. In general, but depending on the relative degrees of σ- and π-bonding in a given complex, the two orbital sets overlap differently with suitable ligand functions. The effect on interelectron repulsion parameters for the different orbitals should be different and might be expressed in terms of a 'differential expansion' of the t_2 and e orbital functions. Our initial discussion of these effects is divided into two main sections. The first concerns the definition and experimental determination of Nephelauxetic factors called $\beta_{33}\beta_{35}$ and β_{55}, while the second relates to various theoretical conclusions which have been derived from these data.

DIFFERENT EXPERIMENTAL NEPHELAUXETIC RATIOS

As discussed in chapter 3, d orbital splittings require at most two interelectron repulsion parameters: F_2 and F_4 or B and C. However, if the weak-field limit is at all inappropriate quantitatively, the Hamiltonian for interelectron repulsion must reflect the true symmetry of the system, rather than the spherical symmetry of the free-ion. In cubic symmetry, for example, we would recognize subshell configurations of the type $t_{2g}^a e_g^b$. Instead of two interelectron repulsion parameters we now require nine, as the following argument[114] shows.

We represent orbitals transforming as t_{2g} (in O_h symmetry) by ξ, η, ζ and those transforming as e_g by u and v. There are fifteen linearly independent two-fold products of these functions; five diagonal and ten off-diagonal, *viz.*

$$\xi^2, \eta^2, \zeta^2, u^2, v^2, \xi\eta, \eta\zeta, \zeta\xi, \xi u, \eta u, \zeta u, \xi v, \eta v, \zeta v, uv. \tag{9.7}$$

Linear combinations of these product wavefunctions which transform as the irreducible representations of O_h may be formed:

$$\begin{aligned}
a_{1g}&: \quad \frac{\xi^2+\eta^2+\zeta^2}{\sqrt{3}} \quad \text{and} \quad \frac{u^2+v^2}{\sqrt{2}},\\
e_g u&: \quad \frac{-\xi^2-\eta^2+2\zeta^2}{\sqrt{6}} \quad \text{and} \quad \frac{u^2-v^2}{\sqrt{2}},\\
e_g v&: \quad \frac{\xi^2-\eta^2}{\sqrt{2}} \quad \text{and} \quad \frac{-uv}{\sqrt{2}},\\
t_{1g}x&: \quad \frac{-\sqrt{3}}{2}\xi u - \tfrac{1}{2}\xi v,\\
t_{1g}y&: \quad \frac{\sqrt{3}}{2}\eta u - \tfrac{1}{2}\eta v,\\
t_{1g}z&: \quad \zeta v,\\
t_{2g}\xi&: \quad \eta\zeta \quad \text{and} \quad -\tfrac{1}{2}\xi u + \frac{\sqrt{3}}{2}\xi v,\\
t_{2g}\eta&: \quad \zeta\xi \quad \text{and} \quad -\tfrac{1}{2}\eta u + \frac{\sqrt{3}}{2}\eta v,\\
t_{2g}\zeta&: \quad \xi\eta \quad \text{and} \quad \zeta u.
\end{aligned} \tag{9.8}$$

For interelectron repulsions, matrix elements of the type $\langle ab|V|cd\rangle$ then require all direct products between these product wavefunctions (9.8) which transform as the totally symmetric representation A_{1g}. There are three from $a_{1g} \times a_{1g}$ (two diagonal and one off-diagonal), three from $e_g \times e_g$, one from $t_{1g} \times t_{1g}$ and three from $t_{2g} \times t_{2g}$, giving ten in all. In octahedral symmetry we thus require ten independent electrostatic $1/r_{ij}$ parameters. One of these, like F_0 in spherical symmetry, may be ignored for the purposes of term *splittings*, leaving nine independent interelectron repulsion parameters to describe nine degrees of freedom.†

Writing,

$$(1) \equiv d_{z^2}, \quad (3) \equiv d_{x^2-y^2}, \quad (4) \equiv d_{xy}, \quad (5) \equiv d_{xz} \quad \text{or} \quad d_{yz},$$

† Nine, for terms arising from a single configuration. If more than one $t_{2g}^a e_g^b$ configuration is involved, Griffith[51] has shown that only one extra (off-diagonal) parameter is required.

TABLE 9.2. *Interelectron repulsion parameters in octahedral symmetry*

Coulomb integrals	Exchange integrals
$J(1,3) \sim A-4B+C$	$K(1,3) \sim 4B+C$
$J(1,4) \sim A-4B+C$	$K(1,4) \sim 4B+C$
$J(3,4) \sim A+4B+C$	$K(3,4) \sim C$
$J(4,4) \sim A+4B+3C$	$K(4,5) \sim 3B+C$
$J(4,5) \sim A-2B+C$	

Jørgensen[c] has constructed the table above of interelectron repulsion integrals in O_h symmetry. The five Coulomb and four exchange integrals in table 9.2 represent the nine independent interelectron repulsion parameters in cubic symmetry. They assume the values indicated in the table under conditions of spherical symmetry.

Jørgensen has also compiled table 9.3 giving excitation energies for

TABLE 9.3. *Excitation energies of some cubic field terms in cubic and spherical approximations*

Con-figuration	Term		Excitation energy with respect to ground term: In cubic symmetry	In spherical symmetry
γ_5^2, t_{2g}^2	$^3\Gamma_4$,	$^3T_{1g}$	0	0
and	$^1\Gamma_5$,	$^1T_{2g}$	$2K(4,5)$	$6B+2C$
γ_5^4, t_{2g}^4	$^1\Gamma_4$,	$^1T_{1g}$	$J(4,4)-J(4,5)$	$6B+2C$
	$^1\Gamma_1$,	$^1A_{1g}$	$J(4,4)-J(4,5)+3K(4,5)$	$15B+5C$
γ_5^3, t_{2g}^3	$^4\Gamma_2$,	$^4A_{2g}$	0	0
and	$^2\Gamma_3$,	^2E_g	$3K(4,5)$	$9B+3C$
$\gamma_3^4\gamma_5^3, t_{2g}^3e_g^4$	$^2\Gamma_4$,	$^2T_{1g}$	$J(4,4)-J(4,5)+K(4,5)$	$9B+3C$
	$^2\Gamma_5$,	$^2T_{2g}$	$J(4,4)-J(4,5)+3K(4,5)$	$15B+5C$
$\gamma_5^3\gamma_3^2, t_{2g}^3e_g^2$	$^6\Gamma_1$,	$^6A_{1g}$	0	0
	$^4\Gamma_1$,	$^4A_{1g}$	$\frac{5}{2}K(1,4)+\frac{5}{2}K(3,4)$	$10B+5C$
	$^4\Gamma_5$,	$^4T_{2g}$	$K(1,4)+K(3,4)+J(4,4)-J(4,5)+K(4,5)$	$13B+5C$
	$^4\Gamma_2$,	$^4A_{2g}$	$4K(1,3)+\frac{3}{2}K(1,4)+\frac{3}{2}K(3,4)$	$22B+7C$
$\gamma_5^6\gamma_3^2, t_{2g}^6e_g^2$	$^3\Gamma_2$,	$^3A_{2g}$	0	0
	$^1\Gamma_3$,	^1E_g	$2K(1,3)$	$8B+2C$
	$^1\Gamma_1$,	$^1A_{1g}$	$4K(1,3)$	$16B+4C$

some cubic-symmetrized configurations in terms of the interelectron repulsion parameters in table 9.2. Terms arising from the strong cubic-field configurations in column 1 of table 9.3 are shown in column 2: both are labelled in Bethe's and in Mulliken's notation. Relative to the ground terms, the energies of these terms due to interelectron repulsion effects are shown in columns 3 and 4. In column 3 the correct cubic symmetry is assumed in which case these excitation energies must be expressed in terms of the nine interelectron repulsion parameters listed in table 9.2. On the other hand if the 'spherical approximation' is made, that is that interelectron repulsion effects will not reflect the lowering of symmetry from spherical to cubic the presence of the ligands makes, the term energies are given by the combination of Racah parameters shown in column 4. The latter figures are obtained from the weak-field term limits of correlation diagrams with which the terms listed in column 2 are correlated.

We should note that a diagonalization of a d configuration matrix under the simultaneous perturbation of interelectron repulsion and crystal fields could be performed with respect to a d orbital basis expressed as $R_{nl}(r).Y_l^m(\theta,\phi)$, i.e. 'weak-field orbitals', or to the strong-field $t_{2g}^a e_g^b$ configurations. The two procedures would give identical results if the 'spherical energies' in column 4 of table 9.3 were used. It is only when both crystal-field and interelectron repulsion Hamiltonians are placed on an equal footing with regard to symmetry, i.e. octahedral rather than octahedral and spherical, that the inexactness of the relationships in table 9.2 becomes apparent, with the resulting proliferation of interelectron repulsion parameters. The question arises, therefore, whether the non-spherical symmetry may be reasonably disregarded or not: such is the subject matter of the following pages.

In extracting experimental values of interelectron repulsion parameters from spectra, the main problem is to remove the effects of, or allow for, crystal-field splittings. In cubic (and occasionally other) symmetry, this separation can often be effected in rather direct and simple ways. Figure 9.1 shows parts of the Tanabe–Sugano diagrams for octahedral d^n complexes ($n = 2$ to 8). These well-known diagrams show the effect of increasing crystal-field strength (here $10Dq$) in O_h symmetry on the various free-ion terms. For $10Dq$ values greater than certain minimum values, some excited levels vary with Dq in a parallel fashion to the appropriate ground terms. These transitions, which are thus nearly independent of Dq, are indicated by arrows in figure 9.1. All transitions shown this way give sharp, 'spin-forbidden'

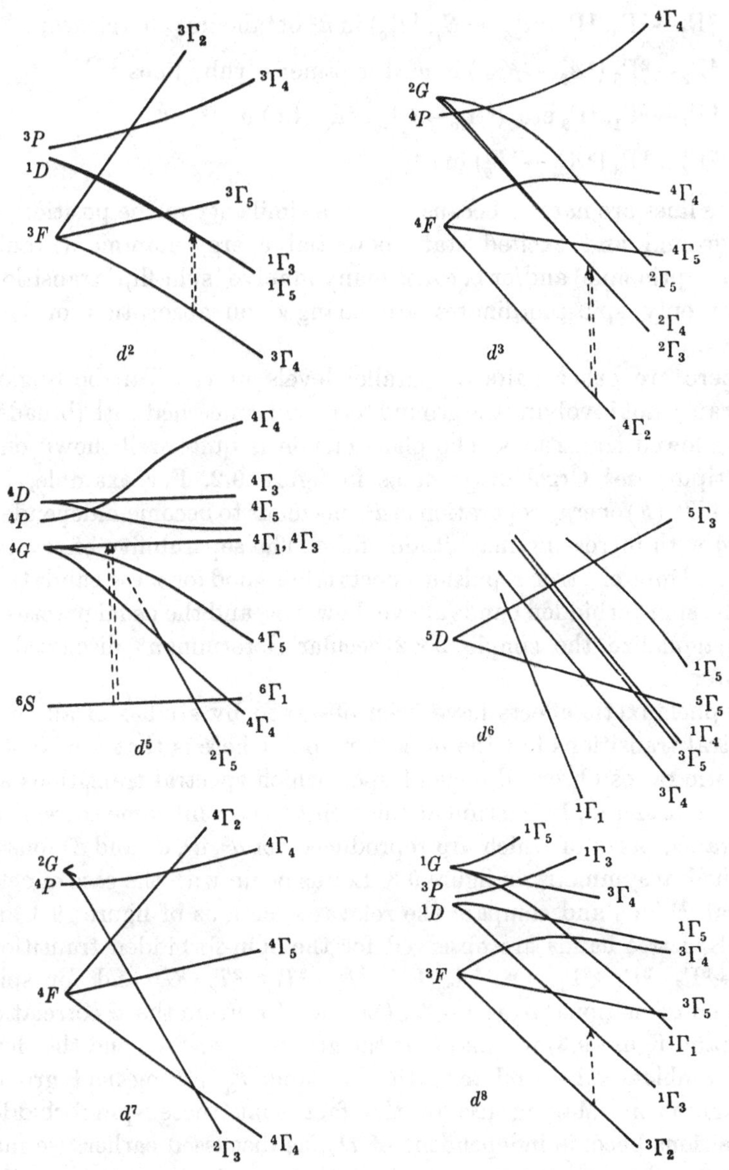

Figure 9.1. Partial Tanabe–Sugano diagrams for d^n systems in O_h symmetry. Spin-forbidden transitions referred to in the text are indicated by arrows.

bands in spectra. We list a few examples:

$^3\Gamma_4 \rightarrow {}^1\Gamma_3, {}^1\Gamma_5$ ($^3T_{1g} \rightarrow {}^1E_g, {}^1T_{2g}$) in d^2 octahedral complexes;

$^4\Gamma_2 \rightarrow {}^2\Gamma_5$ ($^4A_{2g} \rightarrow {}^2T_{2g}$) in d^3, the famous 'ruby lines';

$^6\Gamma_1 \rightarrow {}^4\Gamma_1, {}^4\Gamma_3$ etc. ($^6A_{1g} \rightarrow {}^4A_{1g}, {}^4E_g$ etc.) in d^5;

$^3\Gamma_2 \rightarrow {}^1\Gamma_3$ ($^3A_{2g} \rightarrow {}^1E_g$) in d^8.

[These lines are narrow because of close similarity in the positions of the ground and excited state potential energy minima (Frank-Condon principle) and/or because many involve 'spin-flip' transitions where only spin-coordinates are changed on absorption of light quanta.]

There are other pairs of parallel levels in the Tanabe–Sugano diagrams, not involving the ground term and concerned with (broader) spin-allowed transitions. The phenomenon is quite well shown on a 'multipurpose' Orgel diagram as in figure 9.2. For example, the $^4T_{2g} \leftrightarrow {}^4T_{1g}(F)$ energy separation in d^3 ions tends to become independent of Dq with increasing magnitude of Dq. The separability of crystal-field and interelectron repulsion effects is less good for these bands than for the spin-forbidden bands above, however, and the usual practice is to diagonalize the simple 2×2 secular determinant discussed in chapter 1.

Nephelauxetic effects have been observed by studies of all these spectral transitions but the important point here is that the Nephelauxetic ratios observed depend upon which spectral transitions are being considered. It is useful at this point to consult some correlation diagrams, parts of which are reproduced for d^3, d^5, d^6 and d^8 ions in octahedral symmetry in figure 9.3. Let us begin with the case of octahedral d^3 ions and compare the relevant sections of figures 9.1 and 9.3. Spectral bands are observed for the spin-forbidden transitions $^4\Gamma_2 \rightarrow {}^2\Gamma_3, {}^2\Gamma_4, {}^2\Gamma_5$, i.e. $^4A_{2g}(F) \rightarrow {}^2E_g, {}^2T_{1g}, {}^2T_{2g}({}^2G)$ and for spin-allowed transitions $^4A_{2g}(F) \rightarrow {}^4T_{1g}({}^4F$ and $^4P)$. From the d^3 correlation diagram (figure 9.3) we note that the ground term $^4A_{2g}$ and the three spin-doublets all correlate with the same t_{2g}^3 strong-field ground configuration. Making use of the fact that these spin-forbidden transitions become independent of Dq, as discussed earlier, we may compare their transition energies with those in the corresponding free-ions. The Nephelauxetic effect which is observed is to be associated therefore with transitions involving redistributions within the t_{2g}^3 configuration, as confirmed by the integrals in table 9.3. This effect is represented by the Nephelauxetic ratio β_{55}, the suffices 5 being

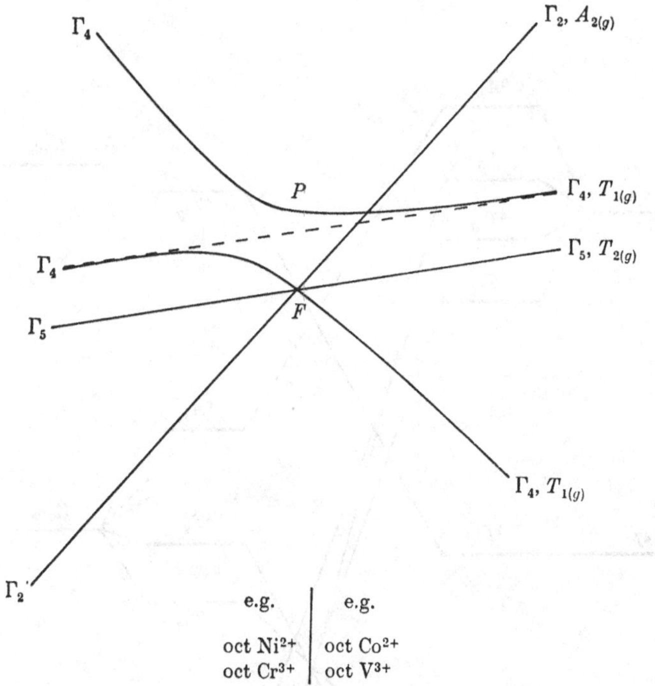

Figure 9.2. Multi-purpose Orgel diagram for spin-allowed transitions in octahedral and tetrahedral d^2, d^3, d^7, d^8 systems.

reminders of the numbering of Bethe's irreducible representations, viz.

$$t_{2g} \equiv \gamma_5, \quad e_g \equiv \gamma_3. \tag{9.9}$$

It should be pointed out here that the arguments above, as those following, are to a degree approximate by virtue of intermediate coupling and configuration mixing effects. While these have been taken into account in the results we shall quote, we are concerned here only to indicate the nature of the parameters β_{55}, etc.: full details are to be found in the references cited at the end of this chapter and further work cited therein.

By contrast, the spin-quartet term ${}^4T_{1g}(F)$ correlates with the first excited configuration $t_{2g}^2 . e_g^1$ (figure 9.3), and the ${}^4T_{1g}(P)$ with $t_{2g}^1 . e_g^2$. The value of B determined by diagonalization of the 2×2 determinant of these ${}^4T_{1g}$ terms for a given complex is then related to the free-ion B_0 value by

$$B = \beta_{35} B_0, \tag{9.10}$$

the mixed suffices representing the change in configuration between ground and excited terms.

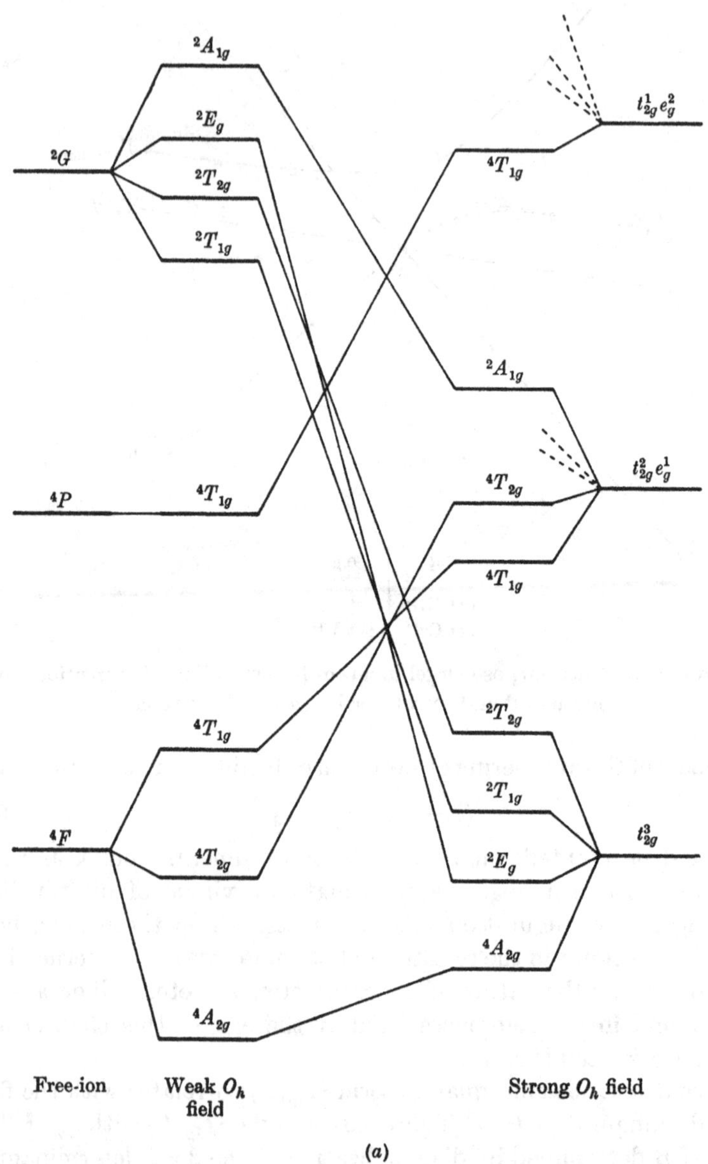

(*a*)

Figure 9.3. Partial correlation diagrams for O_h symmetry, (*a*) for d^3 (*b*) for d^5 (*c*) for d^6 and (*d*) for d^8 configurations.

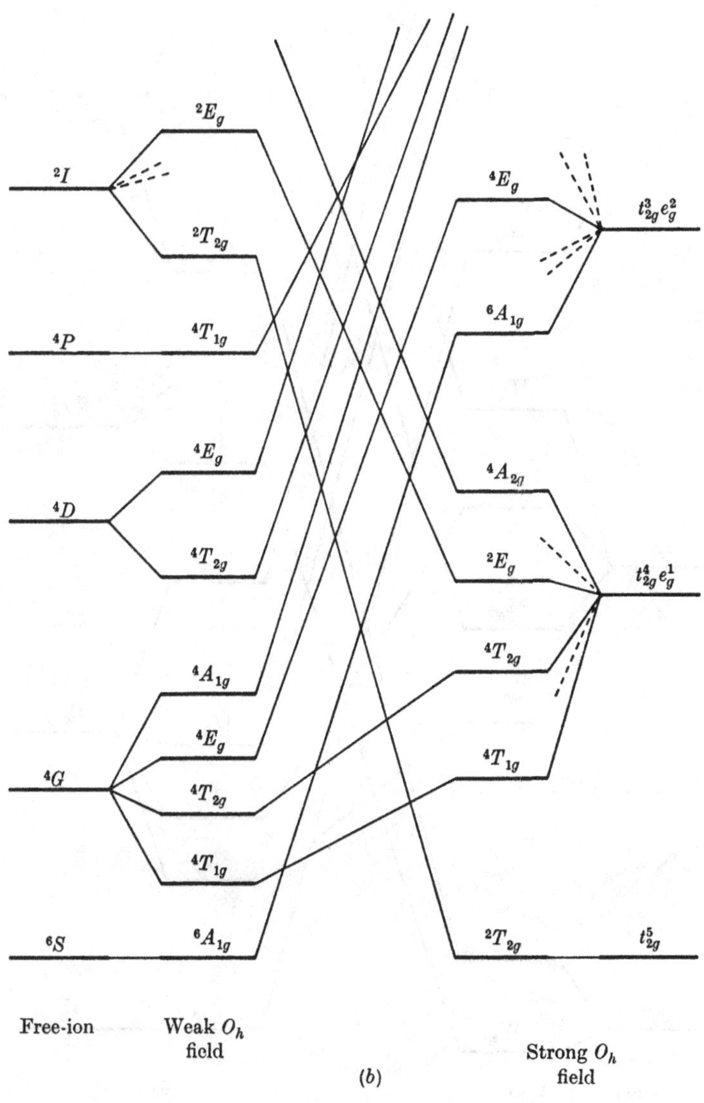

(*b*)

Figure 9.3. (cont.)

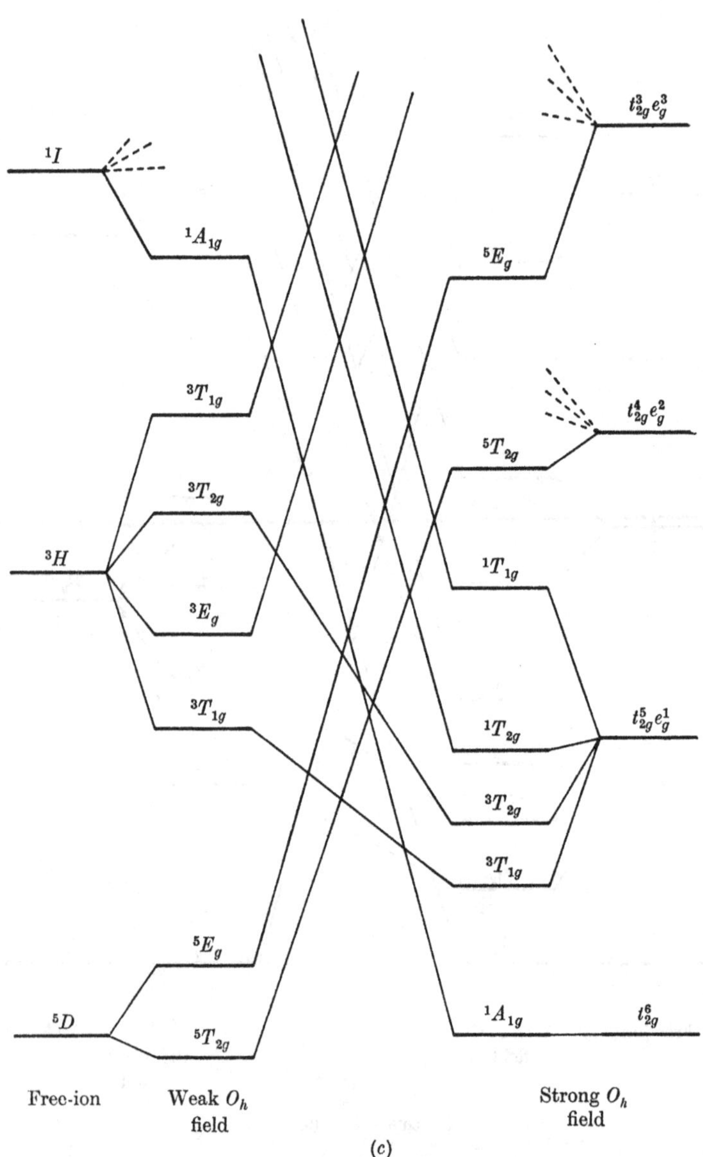

(c)

Figure 9.3. (cont.)

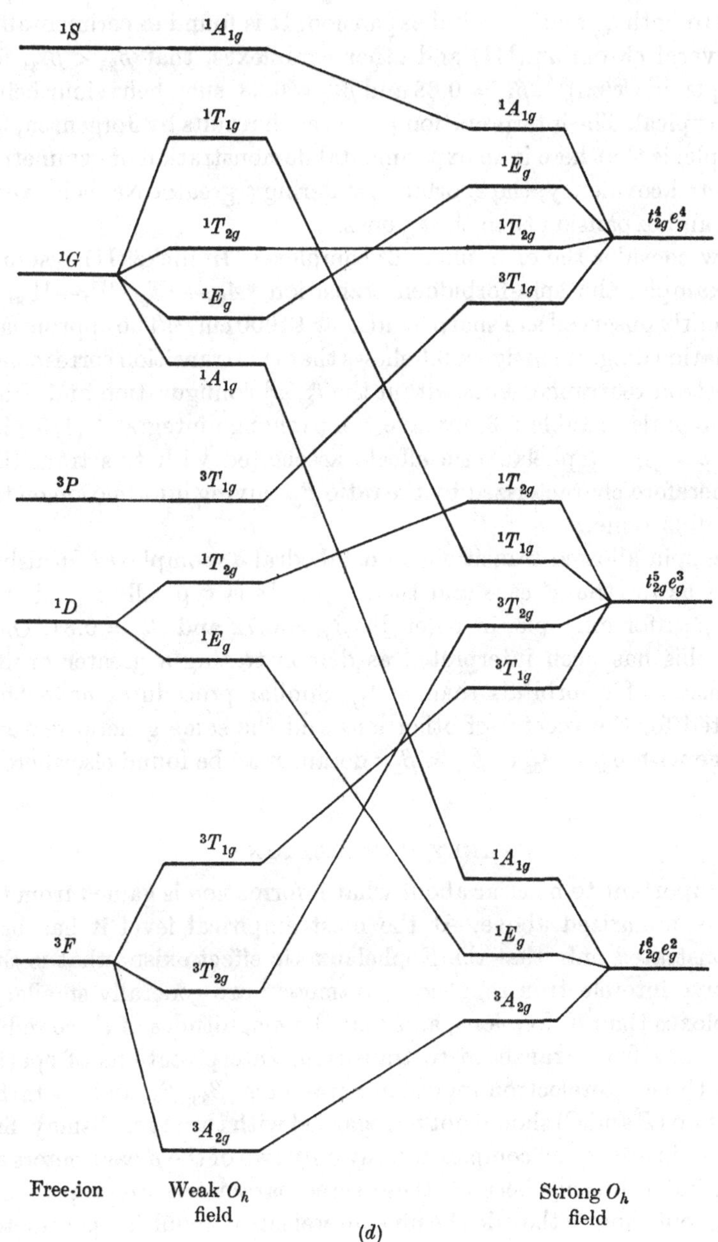

Figure 9.3. (cont.)

Thus β_{55} gives information about t_{2g} orbital expansion while β_{35} refers to both t_{2g} and e_g orbital expansion. It is found experimentally, for several chromium(III) and other complexes, that $\beta_{35} < \beta_{55}$; for example, in $Cr(en)_3^{3+}$, $\beta_{55} = 0.88$ and $\beta_{35} = 0.68$, such behaviour being quite typical. The interpretation put on such results by Jørgensen, for example, is that here is an experimental demonstration of symmetry-restricted covalency, the e_g orbitals suffering a greater Nephelauxetic effect and expansion than the t_{2g} ones.

Now consider the example of d^8 complexes. In nickel(II) systems, for example, the spin-forbidden transition $^3A_{2g} \rightarrow {}^1E_g$ $(^3\Gamma_2 \rightarrow {}^1\Gamma_3)$ is frequently observed as a sharp band near 13000 cm^{-1}. The appropriate correlation diagram in figure 9.3 shows that this transition corresponds to electron rearrangements within the $t_{2g}^6 . e_g^2$ configuration and, from the integrals in table 9.3, involve the exchange integral $K(1, 3)$, i.e. $K(d_{z^2}, d_{x^2-y^2})$. Nephelauxetic effects associated with this transition are therefore characterized by the ratio β_{33}, giving information on the e_g orbitals alone.

The spin-allowed transitions in octahedral d^8 complexes furnish B values as for the d^3 case and thence β_{35}. It is typically found that $\beta_{33} < \beta_{35}$; for example, in $Ni(en)_3^{2+}$, $\beta_{33} = 0.72$ and $\beta_{35} = 0.81$. Once more this has been interpreted as demonstrating a greater orbital expansion of e_g orbitals than of t_{2g}. Similar procedures have been adopted for the spectra of other ions and the same general features emerge with $\beta_{33} < \beta_{35}$ or $\beta_{35} < \beta_{55}$: details may be found elsewhere.[c]

EARLY CONCLUSIONS

It is important to be clear about what information is gained from the work summarized above. At the most empirical level it has been demonstrated only that the Nephelauxetic effect exists, that is that effective interelectron repulsion parameters are generally smaller in complexes than in free-ions, and that the magnitudes of these reductions vary from transition to transition. Interpretations of spectra using three interelectron repulsion parameters β_{33}, β_{35} and β_{55} rather than two (B and C) should not be regarded with too much dismay, first because in any given complex usually only two of the β parameters are used, but primarily because these three parameters are approximations, replacing as they do the nine interelectron repulsion parameters required in cubic symmetry. The theoretical interpretation so far advanced suggests that there are two main causes of these reductions, involving decreased Z_{eff} on the metal on the one hand and electron delocalization into MOs on the other. There are more or less direct pieces of information supporting each of these mechanisms. The classic work of Owen *et al.*[79] on the electron spin resonance spectrum of $IrCl_6^{2-}$ has clearly indicated the migration of unpaired electron spin density from metal to ligand, thus supporting symmetry-restricted covalency as a realistic notion. Evidence for central-field covalency is less direct, but persuasive. Thus crystal-field splittings, or more particularly the integrals G^2, G^4, G^6 (chapter 7) in lanthanide complexes are quite small and perhaps less important in a sense than the Nephelauxetic effects. This result would follow from the usual idea of the $4f$ electrons being buried beneath the electron shells involved in valency. On this basis, the Nephelauxetic effect observed for terms arising from the f^n configuration, presumed to overlap but little with the ligand orbitals, must reflect a change in Z_{eff} rather than electron delocalization. Further the spatially low-lying f orbitals (nearer to the nucleus) will probably be more sensitive to changes in Z_{eff}. These and more detailed considerations discussed later also suggest, in general, that the differences between $\beta_{33}\beta_{35}$ and β_{55} values are best explained by recourse to both mechanisms rather than one. This is the position taken by Jørgensen. As we shall see, others would argue that symmetry-restricted covalency *could* be negligible, or at least that differential expansion of the metal t_{2g} and e_g orbitals could be negligible.

Formalistically,[b] central-field and symmetry-restricted covalency may be separable in the following way. We write a molecular orbital formed between a metal d function and a suitably symmetry-adapted ligand-orbital combination as:

$$\psi_{\text{MO}} = a\psi_d + b\chi_{\text{ligands}}. \tag{9.11}$$

To a first approximation we then have an electron density close to the metal a^2 times as large as in the free-ion, the corresponding interelectron repulsion parameters, being a^4 times as large. The spin–orbit coupling coefficient would be multiplied by a^2. As Jørgensen[b] points out, the contribution to the part of the MO situated near the ligands is neglected in this approximation. While this may not be too serious for spin–orbit coupling parameters which are more 'inner' properties, it is questionable for interelectron repulsion parameters which are more 'outer' ones (see later). This seems particularly so when we recall that the formalism represented by (9.11) refers especially to 'outer', symmetry-restricted covalency. However, we shall continue part way with the argument. The central-field covalency is introduced as a multiplicative factor and we consider,

$$\left.\begin{array}{l} a^2 f_1(Z) \text{ for spin–orbit coupling coefficients,} \\ a^4 f_2(Z) \text{ for interelectron repulsion parameters,} \end{array}\right\} \qquad (9.12)$$

where $f_1(Z)$ and $f_2(Z)$ are functions of the effective nuclear charge, due allowance for the central-field-covalency cloud expansion being implied in Z.

Writing LCAOs for MOs involving t_{2g} and e_g metal orbitals in an octahedral complex, we have:

$$\left.\begin{array}{ll} \gamma_3, e_g: & \psi_3 = a_3\psi_d - b_3\chi_{\text{ligands}}, \\ \gamma_5, t_{2g}: & \psi_5 = a_5\psi_d - b_5\chi_{\text{ligands}}. \end{array}\right\} \qquad (9.13)$$

The Nephelauxetic ratios β_{33} and β_{55} may then be approximated as:

$$\left.\begin{array}{l} \beta_{33} = a_3^4[f_2(Z) \text{ complex}/f_2(Z) \text{ free-ion}], \\ \beta_{55} = a_5^4[f_2(Z) \text{ complex}/f_2(Z) \text{ free-ion}]. \end{array}\right\} \qquad (9.14)$$

If the two electrons being considered are situated one in each subshell, we have:

$$\beta_{35} = a_3^2 a_5^2[f_2(Z) \text{ complex}/f_2(Z) \text{ free-ion}], \qquad (9.15)$$

and, as pointed out by Koide and Pryce:[63]

$$\beta_{33}/\beta_{35} = \beta_{35}/\beta_{55}. \qquad (9.16)$$

Using this formalism, Jørgensen has attempted to quantify the separate roles of central-field (Z) and symmetry-restricted (a) covalency in actual complexes. His conclusions are not unambiguous and partly for this reason and partly in view of what follows we do not attempt to summarize his work here but refer the reader to refs. *b* and *c* at the end of this chapter.

THE VIEWS OF FERGUSON AND WOOD

In some recent publications, Ferguson and Wood[d,e] make two important 'policy' statements. First, that the electrostatic model and 'spherical parameterization' of interelectron repulsion effects should not be abandoned *a priori* and second, that estimates of Nephelauxetic effects should be based on as many spectral transitions as possible for any given complex. We may illustrate their approach with their study of Cr^{3+} ions in ruby, $(Cr/Al)_2O_3$.

Spin-forbidden transitions in ruby, observed by Ferguson and Wood,[d,e] and especially by Kuschida, range 14000–42000 cm^{-1} and are listed in table 9.4. These spectra have been fitted, as shown in the

TABLE 9.4. *Excitation energies* (cm^{-1}) *of doublet terms in ruby*

Observed†	Calculated‡		Calculated §
14447, 14418	^2E	14378	14207
14957, 15168, 15190	$^2T_1$	15049	14817
20993, 21068, 21357	$^2T_2$	20951	21551
29700	$^2A_1$	30160	29960
31000	$^2T_2$	32537	32066
32300	$^2T_1$	32809	32408
34300	^2E	33998	34078
36800	$^2T_1$	37117	37092
40500	$^2T_2$	41132	41607
42300	$^2A_2$	42300	43000

† T. Kuschida, *J. Phys. Soc. Japan* 1966, **21**, 1331.

‡ ^4P at 9600, ^2G at 12160, ^2P at 14500, ^2H at 16350, ^2F at 24300, $^2D_1$ at 15770 cm^{-1}.

§ $B = 650$, $C = 3120\,cm^{-1}$; R. M. Macfarlane, *J. Chem. Phys.* 1963, **39**, 3118.

second column of the table, by diagonalization of the matrices of Finkelstein and Van Vleck. In the weak-field formalism, the energies of the various free-ion terms occur only as diagonal elements, with the exception of the two ^2D terms which arise in d^3 configurations. The relationship $F_4 = 0.07F_2$ was used to fix the values of the three elements corresponding to the energy of the lower ^2D term. These diagonal elements were taken as parameters of the system, each being varied freely and independently. Dq was taken as $1800\,cm^{-1}$, found from the spin-allowed bands in ruby. The term energies under table 9.4 gave the final 'best fit' shown in column 2. Macfarlane had previously fitted these results using the two-parameter model of Racah's B and C parameters as shown in column 3. The fits compare well. At first sight the fact that Macfarlane could fit the spectrum with only two parameters might suggest that Ferguson and Wood's much free fitting had allowed the term energies to vary in a non-productive way. That this is not so is shown nicely by the results in table 9.5. We see that all the empirically determined term energies are very closely reproduced by a two-parameter model using F_2 and F_4 Condon–Shortley interelectron repulsion parameters. For reasons discussed in chapter 3, a Trees correction was included which, in the absence of anything better, was taken to be the same as in Cr^{3+} free-ions: α was fixed at $70\,cm^{-1}$ and was thus not a parameter.

TABLE 9.5. *Term energies in ruby*

Term	Empirical	Calc.†	Term	Empirical	Calc.†
^4P	9600	9500	^2P	14500	14240
^4G	12160	12100	$^2D_1$	15770	16050
^2H	16350	16200			

† $F_2 = 1100$, $F_4 = 84$, $\alpha = 70\,\text{cm}^{-1}$.

In this way, Ferguson and Wood have shown that the complete doublet spectrum in ruby, may be fitted empirically using only two 'spherical' interelectron repulsion parameters. The same conclusion has been reached in the same way for Cr^{3+} in yttrium gallium garnet and in other oxides, in $K_3Cr(CN)_6$ and for Co^{2+} in $ZnAl_2O_4$. The authors imply the result is widespread even if not general. This does not yet mean Jørgensen's β parameters are unnecessary, however, as we shall see.

Quantitatively, the important point made by Ferguson and Wood is that, on complexation, both F_2 and F_4 parameters suffer reduction from their free-ion values but F_2 more than F_4. In ruby, the Nephelauxetic reductions in F_2 and F_4 are 24 % and 18 % respectively. More confirmatory data are quoted in the original paper. An explanation of these results was offered in terms of F_2 being a more 'outer' property than F_4, as follows.

THE CONSEQUENCES OF 'INNER' AND 'OUTER' PROPERTIES

We recall from chapter 3, that a general interelectron repulsion parameter of Condon and Shortley may be written as:

$$F^k = \int_{r_1=0}^{\infty}\int_{r_2=0}^{\infty} \frac{r_<^k}{r_>^{k+1}} R_1^2 . R_2^2 . r_1^2 r_2^2 . dr_1 dr_2. \tag{9.17}$$

These integrals may be broken down, for heuristic reasons, into parts representing contributions to the total integrals for regions inside or outside a certain radius r. Thus, within a sphere of radius r, the Condon–Shortley integrals are given by:

$$F^k(r) = \int_{r_1=0}^{r}\int_{r_2=0}^{r} \frac{r_<^k}{r_>^{k+1}} R_1^2 . R_2^2 . r_1^2 . r_2^2 . dr_1 dr_2. \tag{9.18}$$

It is instructive to compare the values of $F^k(r)$ with their limiting

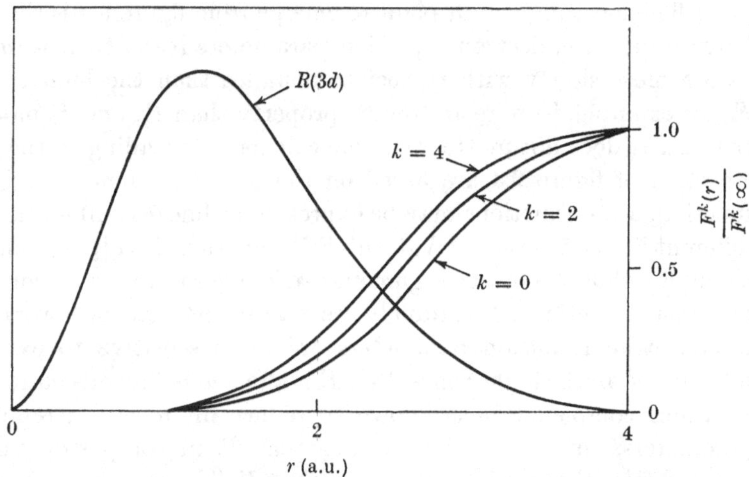

Figure 9.4. Plots of $F^k(r)/F^k(\infty)$ as functions of r compared with $R(3d)$.

values $F^k(\infty)$. Using analytical expressions of Brown and Fitzpatrick,[9] based on the radial wavefunctions R_1 and R_2 being expressed as Slater functions of the form

$$R(3d) = \left[\frac{(2\alpha)^7}{6!}\right]^{\frac{1}{2}} . r^2 e^{-\alpha r}, \tag{9.19}$$

gives

$$F^k(r) = \left[\frac{(2\alpha)^7}{6!}\right]^2 \int_{r_1=0}^{r_1=r} r_1^6 e^{-2\alpha r_1} A \, dr_1, \tag{9.20}$$

where

$$A = \int_{r_2=0}^{r_2=r_1} \frac{r_2^{k+6}}{r_1^{k+1}} . e^{-2\alpha r_2} dr_2 + \int_{r_2=r_1}^{r_2=r} \frac{r_1^k r_2^6}{r_2^{k+1}} . e^{-2\alpha r_2} dr_2.$$

Ferguson and Wood have computed the curves shown in figure 9.4. They make two points about these curves:[d, e]

(1) 'The major contribution to each F^k comes from a region further away from the nucleus than the maximum of the radial distribution.' This seems odd. After all, common sense tells us that interelectron repulsions must maximize where there are the most electrons to repel one another! The point is, of course, that in polar coordinates the elemental volume increases in size with distance from the origin, *viz.*

$$dv = r^2 dr . \sin\theta \, d\theta . d\phi. \tag{9.21}$$

For an electron density radial distribution we plot R^2r^2, which clearly maximizes further from the nucleus than R^2. We came across this same point in chapter 7 in connection with crystal-field radial integrals G^l. Thus a better curve to plot than $R(3d)$ in figure 9.4 would be $R^2(3d)r^2$.

(2) The most important point to emerge from figure 9.4 is that the lower order interelectron repulsion parameters reach their terminal values more slowly with respect to distance than the higher ones. F_2, for example, is more an 'outer' property than F_4, and F_0 more so than F_2. It does not matter that the calculations leading to the construction of figure 9.4 are based on the use of a single-zeta Slater function, as all functions may be expressed as linear combinations of different Slater functions (Watson's SCF functions involving four, for example): thus the simple argument carries over to the more complex or realistic functions. Accordingly, we may expect that the lower order interelectron repulsion parameters are more sensitive to external influence than the higher ones; that F_2 is affected by ligands more than F_4. Since complexation is known to reduce interelectron repulsion parameters, the implication is thus that F_2 in complexes will be reduced from the free-ion value more than F_4. This is what is observed experimentally.

Ferguson and Wood claim[d,e] their analysis 'clearly indicates that differential expansion of e and t_2 orbitals is not important for the calculation of the ligand-field energy levels'. This certainly seems to follow from the preceding arguments. But then what can be said about the consistently different behaviours of β_{33}, β_{35} and β_{55}? Part of the answer might lie in these β ratios being obtained from a single band in each case. Ferguson and Wood point out that 'β_{35} is a measure of the effective energy of the ^4P term in the ligand field, while β_{55} is a measure of the effective doublet term energies, which contribute to the ^2E term, the latter quantity having been obtained by using an approximation. This procedure, based on the energy of one doublet term and the off-diagonal elements of the ^2E strong-field matrix, is too artificial and incomplete for the parameter to have a clear-cut physical meaning'.[d,e] A further point, however, is suggested by the excitation energies listed in table 9.3. Thus, β_{55} values were obtained from d^3 $^4\Gamma_2 \rightarrow {}^2\Gamma_3$, $^2\Gamma_4$, $^2\Gamma_5$ transitions, each of which approximates to a linear combination of Racah B and C parameters in the ratio 3:1. Values for β_{33} were derived from $^3\Gamma_2 \rightarrow {}^1\Gamma_3$ transitions in d^8 involving B and C parameters in the ratio 4:1. Now we recall that

$$B = F_2 - 5F_4 \quad \text{and} \quad C = 35F_4. \tag{9.22}$$

Earlier arguments and experiments give greater percentage reductions in F_2 than F_4 (and $F_2 > F_4$) and hence from (9.22), greater reduction in B than in C (actually very much more so, as we shall see). From the immediately preceding analysis of β values we have β_{33} more dependent on B than is β_{55} (by the ratio 4:3) and so β_{33} values should be less than β_{55}: β_{35} would be intermediate. This was found experimentally. Thus, qualitatively at least, the variation in β values found experimentally corresponds to the general idea of F_2 being more reduced than F_4. It must be admitted straight away that some circular argument is involved here in that the ratios quoted above for the relative participation of B and C in the various β values are based on the assumption of spherical symmetry! Nevertheless, insofar as such an assumption is valid, the above argument shows that an interpretation of spectra in terms of Jørgensen's β values has an equivalence in terms of F_2 and F_4 values. The converse situation is less well demonstrated. As Ferguson and Wood say, 'the use of the Racah formalism disguises the conclusion ...that F_2 and F_4 for the ion in a ligand field are different fractions of the free-ion values and that the differences can be accounted for, qualitatively, without recourse to differential orbital expansion'.[d,e] They do, however, point to a few systems whose spectra do not appear analysable in terms of the two-parameter model discussed above. As they suggest, it is possible that spectroscopic analysis (in most, but perhaps not all, cases) is too insensitive to detect e_g and t_{2g} covalency differences which 'must' [sic] be there.

INTERPRETATIONS II

Ferguson and Wood write about a change in shape of metal orbitals on interaction with the ligands. Thus, the data for ruby, for example, gives the ratio F_2/F_4 as 13.1 as compared with 14.2 in the free-ion. An obvious conclusion would be to suppose that such a result follows from the Nephelauxetic expansion of the metal orbitals due to decreasing Z_{eff} as in the central-field covalency picture. But this change is actually opposite to that in going from Cr^{3+} to Cr^0 as the following simple argument shows.

The Condon–Shortley interelectron repulsion parameters are somewhat similar to the crystal-field G^l integrals we discussed in chapter 7. In each case, we are concerned with the repulsion between electrons expressed by the $1/r_{ij}$ operator. In the case of the crystal-field integrals we viewed the fixed (ligand) electron as setting up a potential represented by potential cusps. We can do the same for interelectron repulsion parameters but here the radial coordinate of the cusp varies in space in the same manner as the metal wavefunctions. We sketch this in figure 9.5 in which the wavefunction $R.r$ represents the radial function of electron 1, say, and the four sets of cusps depict some of the possible coordinates of electron 2. As for the crystal-field radial in-

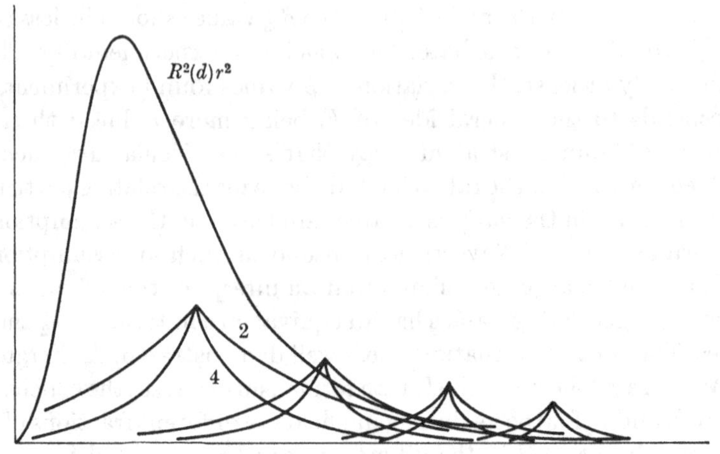

Figure 9.5. Illustrating breakdown of F^k integrals.

tegrals, the upper curve in each cusp represents the potential set up in second order and the lower one, in fourth order. While pointing out this similarity of Cp to F_2 and of Dq to F_4, it is most important not to extrapolate too far. Thus when we enquire how orbital expansion affects interelectron repulsion parameters, we must remember that the cusp distribution representing the 'other' electron alters at the same time. Decreasing Z_{eff} expands the orbitals and increases the average interelectron separation. Interelectron repulsion parameters all decrease. However, F_4 falls off with increasing distance faster than F_2 as may be seen by the cusp representation in figure 9.5 or by the $F^k(r)/F^k(\infty)$ curves in figure 9.4. That is, a decrease in Z_{eff} is accompanied by an increase in the F_2/F_4 ratio, as F_2 is greater than F_4.

Ferguson and Wood therefore argue that, as the observed reduction of the F_2/F_4 ratio on complexation is opposite to the effect associated with a simple reduction in formal oxidation state, 'the radial functions in the ligand field cannot be considered to be the same as free ion functions with a reduced nuclear charge'. This is taken to imply a change in shape relative to the free-ion orbitals. This in turn may be expressed by a variation in coefficients and/or terms in the multi-zeta set of Slater orbitals used to describe these functions with changing Z_{eff}. Ferguson and Wood end their paper concluding that spectroscopic analysis in chromium(III) and some other ions indicates modification of the metal d orbitals in a manner not simply described by a change in Z_{eff}, and that such modification is similar for both t_{2g} and e_g orbitals.

INTERPRETATIONS III

The 'parameterized electrostatic approach' to crystal-field integrals discussed in chapter 7 led to interesting conclusions and a rationale of experimental trends. It is equally interesting to apply this approach to interelectron repulsion parameters and the Nephelauxetic effect. The relations between interelectronic repulsion parameters and Z_{eff} was investigated in connection with figure 9.5 above and may be summarized:

$$\left.\begin{array}{ll}\text{Decreasing } Z_{\text{eff}}: & F_k \text{ decrease, } F_2/F_4 \text{ increases} \\ \text{(orbital expansion)} & \end{array}\right\} \quad (9.23)$$

We would like to discover analogous trends concerned with bond length variation and changes in charge on the ligands, as done for the G^l integrals. For this we recall the work of Craig and Magnusson.[15,16] These authors were principally interested in orbital contraction effects in connection with the availability of certain orbitals in bonding. We shall not reproduce this aspect of their work here but rather adapt the extensions they made of their ideas to repulsive interactions as in crystal-field theory.

We discussed in chapter 2 how the potential set up by a group of ligands could be expanded in a series of spherical harmonics, the largest and leading term being of the spherically symmetric harmonic Y_0^0. Being concerned with crystal-field orbital splittings only, we paid no further attention to this term which affects all $(n)d$ orbital energies equally. It is, however, a most important term in connection with interelectron repulsion energies and indeed our discussion here initially considers this term only, leaving the effects of higher terms (Y_4^m in O_h symmetry, for example) as corrections to be considered later. In this 'zeroth-order' approximation, the negatively charged (point-charge or point-dipole) ligands are represented by a uniform, negatively charged, spherical shell. As known from a standard theorem in electrostatics, the potential *inside* such a sphere is uniform (for six ligands of charge ze, distant a from the metal, the potential is $6ze/a$). Thus no work is done by moving an electron from one point to another within the spherical shell. Even though the energy, as described by the ionization potential for example, of such electrons would be altered, the form of wavefunctions inside the shell would not. However, the potential seen by any electron outside the shell is equivalent to that experienced if the charge on the shell were accumulated at its centre, and decreases with distance. Outer electrons thus experience a smaller effective nuclear charge than inner ones and so the outer parts of the wavefunctions are expanded. Note that this electrostatic expansion effect is additional to a central-field covalency caused by transfer of charge from ligand to metal. For a (d) wavefunction which is partly inside and partly outside the shell, it is not quite correct to conclude that the part inside remains unchanged and the part outside expands. This is because the inner and outer parts of the wavefunction are 'connected' by the kinetic energy of the electrons they contain, there being a balancing between the increased potential energy and decreased kinetic energy. As a result, both parts expand in general, but the outer part much more so. Craig and Magnusson have made approximate calculations of the magnitude of these effects and we shall comment on these later. However, for the moment we consider the effect in a qualitative fashion, in much the same way as we did for the G^l integral calculations of Ballhausen and Ancmon[4] in chapter 7.

In this model, increasing bond length corresponds to increasing radius of the spherical shell. The larger the radius of the shell, the smaller is the part of the d-orbital lying outside it and hence the smaller is the net orbital expansion. Now as figure 9.4 showed, F_2 is more sensitive to outer perturbation than is F_4. This may be expressed in another way, as shown in figure 9.6 which shows how, for inner parts of the wavefunction, the ratio between lower- and higher-order interelectron repulsion integrals is smaller than the final value: F^4 reaches 90 %, say, of its final value faster than F^2. If we consider for the moment the extremes of inhomogeneous orbital expansion in favour of the outer parts as represented by an infinite expansion (i.e. zero interelectron repulsion effects) outside the spherical shell, with no expansion within it, we can see the net effect is to reduce all F_k, but especially the lower order F_k. The inhomogeneous expansion thus tends to reduce F_2/F_4. However, the generally greater electron–electron separations implied by an expansion tend, as before, to reduce F_k and *increase* F_2/F_4. There is, accordingly, some conflict in predictions depending on the degree of inhomogeneity of the orbital expansion. We might summarize then:

$$\text{Decreasing bond length:} \quad F_k \text{ decrease, } F_2/F_4 \text{ little changed.} \tag{9.24}$$

The effective charge on the ligand is the other variable we considered for the G^l integrals. In the present model, increasing ligand charge increases the charge on the spherical shell and so, qualitatively, acts in the same direction as decreasing bond lengths.

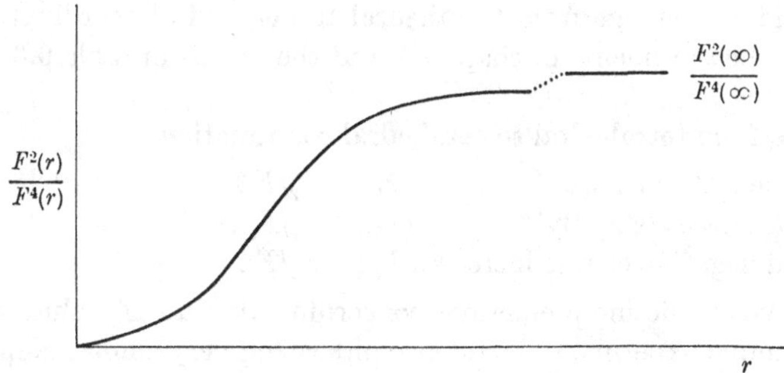

Figure 9.6. Variation of $F^2(r)/F^4(r)$ with r.

SUMMARY OF TRENDS

In table 9.6, we summarize the trends in interelectron repulsion parameters and their ratios as functions of central-field covalency, bond lengths and ligand charges. We represent increasing and decreasing properties by upward and downward pointing arrows, respectively. Already, we see how, at this qualitative level, predictions about F_2/F_4 ratios cannot easily be made by this electrostatic model. It is nevertheless instructive to apply the trends in table 9.6 as far as immediately possible.

TABLE 9.6

I	Decrease bond length	$F_k\downarrow$	F_2/F_4?
II	Transfer negative charge to metal (i.e. decrease Z_{eff})	$F_k\downarrow$	$F_2/F_4\uparrow$
III	Transfer negative charge from ligands	$F_k\uparrow$	F_2/F_4?

(1) We consider the formation of a complex from a free-ion. Relative to that free-ion, three changes may be considered to take place, their effects on interelectron repulsion parameters being as follows:

(*a*) shorter bond lengths (i.e. less than ∞) $F_k\downarrow$ F_2/F_4?
(*b*) decreasing Z_{eff} $F_k\downarrow$ $F_2/F_4\uparrow$
(*c*) increasing ligand charge (i.e. $ze > 0$) $F_k\downarrow$ F_2/F_4?

As in chapter 7 we must take note only of predictions without conflict if we are to maintain the utility of a qualitative approach. All three trends above predict reduction in F_k values: this is observed experimentally.

(2) Consider a change from tetrahedral to octahedral coordination. Re-using the arguments in chapter 7 and the trends in table 9.6, we have:

Change from tetrahedral to octahedral coordination:

(*a*) bond lengths increase $F_k\uparrow$ F_2/F_4?
(*b*) Z_{eff} decreases (slightly)? [$F_k\downarrow$ $F_2/F_4\downarrow$]
(*c*) ligand negative charge increases $F_k\downarrow$ F_2/F_4?

We observe conflicting predictions concerning absolute F_k values and there is ample experimental evidence indicating very similar Nephelauxetic effects in tetrahedral and octahedral geometries. Jørgensen remarks that 'there is no indication of any intrinsically *much* stronger covalent bonding in the tetrahedral chromophores'. In view of the many contributing factors to the Nephelauxetic effect, including configuration interactions on the metal, it is possible that a more cautious statement still would be appropriate.

(3) Table 9.7 lists some spectroscopic and crystallographic data for a short series of chromium(III) complexes, compiled by Wood.[118] For illustration, consider the compounds K_2NaCrF_6 and $CrBr_3$ involving octahedral coordination of chromium(III) ions by fluorine and bromine ligands. We construct the following summary:

Change from CrF_6 to $CrBr_6$:

$$\left.\begin{array}{lll} (a)\ \text{longer bonds} & F_k\uparrow & F_2/F_4? \\ (b)\ \text{decreasing } Z_{eff} & F_k\downarrow & F_2/F_4\uparrow \\ (c)\ \text{decreasing ligand charge} & F_k\uparrow & F_2/F_4? \end{array}\right\} \tag{9.25}$$

TABLE 9.7. *Spectroscopic and crystallographic data for some chromium(III) complexes*

	Free-ion	Ruby	K_2NaCrF_6	CrF_3	$CrCl_3$	$CrBr_3$
Dq(cm^{-1})	0	1630	1610	1460	1370	1340
B(cm^{-1})	920	640	760	740	550	370
C(cm^{-1})	3680	3300	3020	—	3400	3700
C/B	4.0	5.15	4.0	—	6.3	10.0
F_2(cm^{-1})	1446†	1112	1192	—	1036	900
F_4(cm^{-1})	105†	94.5	86.2	—	97.1	106
F_2/F_4	13.8	11.8	13.8	—	10.6	8.5
Cr-ligand distance (Å)		1.90	1.90	1.90	2.38	2.54

† These values disagree slightly with those in table 9.5 due to neglect of Trees' correction in the present case.

The series of compounds in table 9.7, exemplified by the $CrF_6/CrBr_6$ comparison, sets an acid test for the utility of an electrostatic approach to interelectronic repulsion integrals, illustrates some interesting trends and poses some fundamental questions.

As summarized in (9.25) no firm predictions concerning either F_k values or F_k ratios can be made from the present model. This is disappointing in view of the facts we wish to 'explain', *viz.*

Change from CrF_6 to $CrBr_6$:

(1) B decreases (markedly)
(2) C increases
(3) F_2 decreases
(4) F_4 increases
(5) C/B increases (markedly)
(6) F_2/F_4 decreases

Let us note straightaway that if a Nephelauxetic series is constructed by reference to B values [as in (9.2)], we observe the 'normal' series:

$$K_2NaCrF_6 > \text{ruby} > CrCl_3 > CrBr_3.$$

However, the series is *exactly reversed* for trends in C. Similar behaviours with respect to F_2 and F_4 are also observed. Conclusions concerning the extent of covalency which are based on the normal Nephelauxetic series may deserve a second look, if they depend on no more than interpolation.

At a pedantic level, one could say the conflicting predictions about F_k values in (9.25) concur with experimental fact, in that F_4 values increase while F_2 values decrease on replacing F by Br. However, as for our discussions of crystal-field radial integrals, the only purpose for, or defence of, a parameterized electrostatic model is on the grounds of its utility. So far, for interelectronic repulsion parameters it is found wanting in this most important respect. It is premature yet to discard the philosophy of the electrostatic approach to interelectronic repulsion parameters, however, as we now discuss.

The mechanisms and trends listed in table 9.6 are appropriate only for a spherically-symmetric system, qualitatively because cubic symmetry requires more parameters than the Condon–Shortley F_k, but quantitatively because the model hitherto has ignored the higher-order crystal-field terms in the Craig and Magnusson treatment.

The concept of inhomogeneous orbital-expansion within a given orbital was discussed earlier by reference to the leading term in the expansion of the perturbing ligand potential. All discussions referred to a 'spherical shell' of charge. Craig and Magnusson, however, also considered the effects of higher terms in the expansion. As would be expected, relaxation of the spherical symmetry does lead to different results for different orbitals. In the case of orbital expansion resulting from the effect of six negative ligand charges in octahedral array around the metal, Craig and Magnusson[15,16] demonstrated greater expansion of the e_g set than the t_{2g} set,† the difference being most significant for shorter bond lengths and lower Z_{eff} values. Now a greater expansion of e_g orbitals relative to t_{2g} implies $\beta_{33} < \beta_{35} < \beta_{55}$, which in turn implies F_2/F_4 is less than in the corresponding spherical system (see earlier). However, as differential orbital expansion increases (as opposed to inhomogeneous expansion), the 'spherical parameters' F_2 and F_4 (or B and C) gradually lose some meaning. Even so the work of Craig and Magnusson appears to suggest that shorter bond lengths and/or lower Z_{eff} values increase the size of e_g orbitals relative to t_{2g} ones, corresponding roughly to a decrease in the F_2/F_4 ratio, insofar as these parameters remain useful.

The change from CrF_6 to $CrBr_6$ involves longer bonds and decreasing Z_{eff} and so we have conflict in this respect as to whether there will be a significant differential orbital expansion. The high formal charge on the metal in these systems might, in any case, reduce the importance of differential orbital expansion.

It does not appear possible, therefore, to make simple correlations between observed interelectron repulsion parameters and a 'parameterized crystal-field model', but before leaving this avenue of thought, it seems worth while making some general observations. The trends we have discussed are qualitative only. Not only may the sensitivity of an F_2/F_4 ratio to any mechanism differ from the sensitivity of absolute F_k values, but also relative sensitivities of interelectron repulsion parameters to the different mechanisms may vary. Further, these gradients may be non-linear. For example, in a complex involving a high formal metal oxidation state, the metal wavefunctions may be sufficiently contracted not to extend significantly outside the charged shell (spherical or otherwise) representing the ligands. In such circumstances, inhomogeneous orbital expansion effects would diminish in importance, leaving factor II, determined by 'central-field covalency' effects, as the most important mechanism. This result

† Craig and Magnusson[15,16] point out that there is a lack of symmetry with respect to the sign of the perturbation. Thus, for positive ligand charges, producing orbital contraction the differential effect on various orbitals is very small, if not negligible.

would be augmented by the fact that mechanism II itself would be most sensitive under these same conditions, any decrease in Z_{eff} being most efficacious in the highly contracted, 'over-crowded' metal orbitals. If this sort of argument is applied to the chromium(III) complexes above, the experimental trend in F_2/F_4 ratios is still not reproduced, factor (*b*) in (9.25) predominating.

We have witnessed the inability of the parameterized electrostatic approach to reproduce observed trends in interelectronic repulsion parameters and this raises some important questions.

(1) Does the experimental trend in the F_2/F_4 values in table 9.7 prove that 'spherical F_k parameters are inappropriate and that differential orbital expansion – the electrostatic version of symmetry-restricted covalence – is significant?

(2) Does the failure to reproduce interelectron repulsion parameters imply the success in reproducing crystal-field parameters was specious?

We can comment on the questions rather than actually answer them. A single observation serves to diminish the concern both questions raise. The fundamental difference between interelectron repulsion and crystal-field parameters, repeated so many times already, is that the former involve two non-localized quantum particles while the latter involves one quantum particle (at a time, that is) and one fixed classical ligand charge. Because of this most fundamental distinction, the *shape* of the orbitals we discuss is most important for interelectron repulsion parameters, at least as far as F_2/F_4 *ratios* are concerned: crystal-field parameters are involved more directly with orbital extension, bond lengths and ligand charges. It might be reasonable to expect our approach above to correlate with trends in F_k but not their ratios. This much appears to work, insofar that the predicted trends will usually be in conflict [e.g. (9.25) above], a lack of unanimity in these trends cancels any predictive value the approach would have.

The fact that arguments based on the work of Craig and Magnusson would lead us to expect differential $t_{2g} - e_g$ orbital expansion effects to be minimized in chromium(III) complexes relative, say, to cobalt(II) or nickel(II), makes us reluctant to deduce that the failure to explain the F_2/F_4 ratios in table 9.7 implies considerable differential orbital expansion. There remains, however, the purely molecular-orbital approach which might well lead to greater symmetry restricted covalency effects in the bromide relative to the fluoride. As for the remaining question of which of Jørgensen's or Ferguson and Wood's approach to interelectron repulsion parameters is the more appropriate, we can add little to what has already been discussed. In that an

F_2/F_4 ratio less than the free-ion value would correspond to the inequality

$$\beta_{33} < \beta_{35} < \beta_{55},$$

even in perfect spherical symmetry, it appears that a detailed demonstration of differential $t_{2g} - e_g$ orbital expansion, or lack of it, is yet to be made.

GENERAL REFERENCES

(*a*) C. K. Jørgensen, *Absorption Spectra and Chemical Bonding in Complexes*, Pergamon Press, Oxford, 1962.

(*b*) C. K. Jørgensen, *Modern Aspects of Ligand Field Theory*, North-Holland, Amsterdam, 1971.

(*c*) C. K. Jørgensen, The Nephalauxetic Series, *Prog. Inorg. Chem.* 1962, **4**, 73.

(*d*) J. Ferguson, Spectroscopy of 3*d* complexes, *Prog. Inorg. Chem.* 1970, **12**, 159.

(*e*) J. Ferguson and D. L. Wood, *Aust. J. Chem.* 1970, **23**, 861.

REFERENCES

References indicated by a letter in the text are to be found at the end of the relevant chapter.

1. W. A. Baker and M. G. Phillips, *Inorg. Chem.* 1966, **5**, 1042.
2. C. J. Ballhausen, *Introduction to Ligand Field Theory*, McGraw-Hill, New York, 1962.
3. C. J. Ballhausen, *Introduction to Ligand Field Theory*, McGraw-Hill, New York, 1962, p. 54.
4. C. J. Ballhausen and E. M. Ancmon, *Mat. Fys. Medd. Dan. Vid. Selsk.* 1958, **31**, 2.
5. C. J. Ballhausen and C. R. Hare, *J. Chem. Phys.* 1964, **40**, 788 and 792.
6. H. D. Bedon, S. M. Horner and S. Y. Tyree Jun., *Inorg. Chem.* 1964, **3**, 647.
7. H. A. Bethe, *Ann. der Physik*, 1929, **3**, 133.
8. D. M. Brink and G. R. Satchler, *Angular Momentum*, Clarendon Press, Oxford, 1968.
9. D. A. Brown and N. J. Fitzpatrick, *J. Chem. Soc.* (*A*) 1966, 941.
10. C. D. Burbridge, D. M. L. Goodgame and M. Goodgame, *J. Chem. Soc.* (*A*) 1967, 349.
11. A. O. Caride, H. Panepucci and S. I. Zanethe, *J. Chem. Phys.* 1971, **55**, 3651.
12. E. Cartmell and G. W. A. Fowles, *Valency and Molecular Structure*, Butterworths, London, 1956.
13. M. Ciampolini, *Inorg. Chem.* 1966, **5**, 35.
14. D. P. Craig, A. Maccoll, R. S. Nyholm, L. E. Orgel and L. E. Sutton, *J. Chem. Soc.* 1954, 332.
15. D. P. Craig and E. A. Magnusson, *J. Chem. Soc.* 1956, 4895.
16. D. P. Craig and E. A. Magnusson, *Disc. Farad. Soc.* 1958, **26**, 116.
17. F. A. Cotton, *Chemical Applications of Group Theory*, Interscience, New York, 1963.
18. T. W. Couch and G. P. Smith, *J. Chem. Phys.* 1970, **53**, 1336.
19. P. Day and C. K. Jørgensen, *J. Chem. Soc.* 1964, 6226.
20. L. Dubicki, M. A. Hitchman and P. Day, *Inorg. Chem.* 1970, **9**, 188.
21. J. D. Dunitz and L. E. Orgel, *J. Chem. Phys.* 1955, **23**, 954.
22. D. E. Ellis, A. J. Freeman and P. Ros, *Phys. Rev.* 1968, **176**, 688.
23. H. Eyring, J. Walter and G. E. Kimball, *Quantum Chemistry*, Wiley, New York, 1944.
24. J. Ferguson, *J. Chem. Phys.* 1964, **40**, 3406.
25. J. Ferguson, *Prog. Inorg. Chem.* 1970, **12**, 159.
26. J. Ferguson and D. L. Wood, *Aust. J. Chem.* 1970, **23**, 861.
27. R. P. Feynman, R. B. Leighton and M. Sands, *The Feynman Lectures on Physics*, Vol. 3, Addison-Wesley, 1965.
28. B. N. Figgis, M. Gerloch, J. Lewis and R. C. Slade, *J. Chem. Soc.* (*A*) 1968, 2028.
29. B. N. Figgis, M. Gerloch and R. Mason, *Acta Cryst.* 1964, **17**, 506.
30. A. J. Freeman and R. E. Watson, *Phys. Rev.* 1960, **120**, 1254.
31. C. D. Garner and F. E. Mabbs, *J. Chem. Soc.* (*A*) 1970, 1711.

32. M. Gerloch, *J. Chem. Soc.* (*A*) 1968, 2023.
33. M. Gerloch, J. Kohl, J. Lewis and W. Urland, *J. Chem. Soc.* (*A*) 1970, 3269.
34. M. Gerloch, J. Kohl, J. Lewis and W. Urland, *J. Chem. Soc.* (*A*) 1970, 3283.
35. M. Gerloch and J. Lewis, *Rev. Chim. Min.* 1969, **6**, 19.
36. M. Gerloch, J. Lewis, G. G. Phillips and P. N. Quested, *J. Chem. Soc.* (*A*) 1970, 1941.
37. M. Gerloch, J. Lewis and R. Rickards, *J. Chem. Soc., Dalton*, 1972, 980.
38. M. Gerloch, J. Lewis and W. R. Smail, *J. Chem. Soc.* (*A*) 1971, 2434.
39. M. Gerloch, J. Lewis and W. R. Smail, *J. Chem. Soc., Dalton*, 1972, 1559.
40. M. Gerloch and D. J. Mackey, *J. Chem. Soc.* (*A*) 1970, 3040.
41. M. Gerloch and J. R. Miller, *Prog. Inorg. Chem.* 1968, **10**, 1.
42. M. Gerloch and P. N. Quested, *J. Chem. Soc.* (*A*) 1971, 3729.
43. M. Gerloch, P. N. Quested and R. C. Slade, *J. Chem. Soc.* (*A*) 1971, 3740.
44. M. Gerloch and R. C. Slade, *J. Chem. Soc.* (*A*) 1969, 1012.
45. M. Gerloch and R. C. Slade, *J. Chem. Soc.* (*A*) 1969, 1022.
46. M. Gerloch, R. C. Slade and W. R. Smail, *in preparation.*
47. H. M. Gladney and A. Veillard, *Phys. Rev.* 1969, **180**, 385.
48. K. I. Gondaira, *J. Phys. Soc. Japan* 1966, **21**, 933.
49. D. M. L. Goodgame, M. Goodgame, M. A. Hitchman and M. J. Weeks, *Inorg. Chem.* 1966, **5**, 635.
50. H. B. Gray, in *Electrons and Chemical Bonding*, Benjamin, New York, 1965, p. 198.
51. J. S. Griffith, *Theory of Transition-Metal Ions*, Cambridge University Press, 1961.
52. J. S. Griffith, *J. Chem. Phys.* 1964, **41**, 576.
53. M. A. Hitchman, *J. Chem. Soc., Faraday II*, 1972, **68**, 846.
54. J. Hubbard, D. E. Rimmer and F. R. A. Hopgood, *Proc. Phys. Soc.* 1966, **88**, 13.
55. J. P. Jesson, *J. Chem. Phys.* 1968, **48**, 161.
56. B. R. Judd, *Operator Techniques in Atomic Spectroscopy*, McGraw-Hill, New York, 1962.
57. C. K. Jørgensen, *Modern Aspects of Ligand Field Theory*, North Holland, Amsterdam, 1971.
58. C. K. Jørgensen, *Prog. Inorg. Chem.* 1962, **4**, 73.
59. C. K. Jørgensen, R. Pappalardo and H. H. Schmidtke, *J. Chem. Phys.* 1963, **39**, 1422.
60. B. L. Kalman and J. W. Richardson, *J. Chem. Phys.* 1971, **55**, 4443.
61. S. F. A. Kettle, *Inorg. Chem.* 1965, **4**, 1821.
62. W. H. Kleiner, *J. Chem. Phys.* 1952, **20**, 1784.
63. S. Koide and M. L. H. Pryce, *Phil. Mag.* 1958, **3**, 607.
64. M. Kotani, A. Amemiya, E. Ishiguro and J. Kimura, *Tables of Molecular Integrals*, Maruzen, 1955.
65. D. R. Layzer, *Dissertation*, Harvard University, 1950 (unpublished).
66. A. B. P. Lever, *Coord. Chem. Rev.* 1968, **3**, 119.
67. A. B. P. Lever, B. R. Hollebone and J. C. Donini, *J. Amer. Chem. Soc.* 1971, **93**, 6455.
68. O. Matsuoka, *J. Phys. Soc. Japan* 1970, **28**, 1296.
69. J. W. Moskowitz, C. Hollister, C. J. Hornback and H. Basch, *J. Chem. Phys.* 1970, **53**, 2570.
70. D. S. McClure, in *Advances in the Chemistry of Coordination Compounds*, ed. S. Kirshner, MacMillan, New York, 1961, p. 498.

71. R. S. Mulliken, *Phys. Rev.* 1932–3; **40**, 55; **41**, 49, 751; **43**, 279, and *J. Chem. Phys.* 1933, **1**, 492; 1935, **3**, 375, 506.
72. R. S. Mulliken, *J. Chim. Phys.* 1949, **46**, 497.
73. R. S. Mulliken, C. A. Rieke, D. Orloff and H. Orloff, *J. Chem. Phys.* 1949, **17**, 1248.
74. J. N. Murrell, S. F. A. Kettle and J. M. Tedder, *Valence Theory*, Wiley, New York, 1965.
75. M. J. Norgett, J. H. M. Thornley and L. M. Venanzi, *J. Chem. Soc.* (*A*) 1967, 540.
76. P. O'D. Offenhartz, *J. Chem. Phys.* 1967, **47**, 2951.
77. P. O'D. Offenhartz, *J. Amer. Chem. Soc.* 1969, **91**, 5699.
78. L. E. Orgel, *An Introduction to Transition Metal Chemistry*, Methuen, London, 1960.
79. J. Owen, J. H. E. Griffiths and I. M. Ward, *Proc. Roy. Soc.* (*A*) 1953, **219**, 526.
80. R. Pariser and R. G. Parr, *J. Chem. Phys.* 1953, **21**, 466 and 767.
81. L. Pauling, *Nature of the Chemical Bond*, Cornell University Press, New York, 1945.
82. J. R. Perumareddi, *Coord. Chem. Rev.* 1969, **4**, 73.
83. J. C. Phillips, *J. Phys. Chem. Solids* 1959, **11**, 226.
84. T. S. Piper and A. G. Karipides, *Inorg. Chem.* 1962, **1**, 970.
85. D. Polder, *Physica*, 1942, **9**, 709.
86. J. A. Pople, *Trans. Farad. Soc.* 1953, **49**, 1375.
87. J. A. Pople and G. A. Segal, *J. Chem. Phys.* 1965, **43**, 5136.
88. G. Racah, *Phys. Rev.* 1952, **85**, 381.
89. J. W. Richardson, W. C. Nieuwpoort, R. R. Powell and W. F. Edgell, *J. Chem. Phys.* 1962, **36**, 1057.
90. J. W. Richardson, D. M. Vaught, T. F. Soules and R. R. Powell, *J. Chem. Phys.* 1969, **50**, 3633.
91. C. W. Reimann, *J. Phys. Chem.* 1970, **74**, 561.
92. C. C. J. Roothaan, *Rev. Mod. Phys.* 1951, **23**, 69.
93. D. A. Rowley and R. S. Drago, *Inorg. Chem.* 1967, **6**, 1092.
94. C. E. Schäffer, (*a*) *Structure and Bonding*, 1968, 5, 68, (*b*) *Proc. Roy. Soc.* 1967, A**297**, 96.
95. C. E. Schäffer and C. K. Jørgensen, *Mol. Phys.* 1965, **9**, 401.
96. H. H. Schmidtke, *Z. Naturforsch.* 1964, **19***a*, 1502.
97. S. Sugano and R. G. Shulman, *Phys. Rev.* 1963, **130**, 517.
98. D. W. Smith, *J. Chem. Soc.* (*A*) 1969, 1708 and 2529.
99. D. W. Smith, *J. Chem. Soc.* (*A*) 1970, 176 and 2900.
100. D. W. Smith, *Inorg. Chim. Acta* 1971, **5**, 231.
101. D. W. Smith, *J. Chem. Soc.* (*A*) 1971, 1024.
102. D. W. Smith, *J. Chem. Soc.* (*A*) 1971, 1209.
103. D. W. Smith, *J. Chem. Phys.* 1969, **50**, 2784.
104. J. C. Slater, *Quantum Theory of Atomic Structure* Vol. 1, McGraw-Hill, New York, 1960.
105. D. R. Stephens and H. G. Drickamer, *J. Chem. Phys.* 1961, **35**, 427 and 429.
106. K. W. H. Stevens, *Proc. Roy. Soc.* (*A*), 1953, **219**, 542.
107. A. Streitwieser Jun., *Molecular Orbital Theory for Organic Chemists*, Wiley, New York, 1961.
108. Y. Tanabe and S. Sugano, *J. Phys. Soc. Japan*, 1956, **11**, 864.
109. R. E. Trees, *Phys. Rev.* 1951, **83**, 756 and 1951, **84**, 1089.
110. C. W. Ufford and H. B. Callen, *Phys. Rev.* 1958, **110**, 1352.

111. J. H. Van Vleck, *J. Chem. Phys.* 1935, **3**, 803 and 807.
112. J. H. Van Vleck, *Phys. Rev.* 1932, **41**, 208 and *Disc. Farad. Soc.* 1958, **26**, 96.
113. J. H. Van Vleck, *J. Chem. Phys.* 1939, **7**, 72.
114. H. Watanabe, *Operator Methods in Ligand Field Theory*, Prentice-Hall, New Jersey, 1966.
115. R. E. Watson, *Phys. Rev.* 1960, **118**, 1036.
116. R. E. Watson and A. J. Freeman, *Phys. Rev.* 1964, **134**, A 1526.
117. M. Wolfsberg and L. Helmholz, *J. Chem. Phys.* 1952, **20**, 837.
118. D. L. Wood, in *Optical Properties of Solids*, Plenum Press, 1969, chapter 19.
119. S. Yanagawa, *J. Phys. Soc. Japan.* 1955, **10**, 1029.

INDEX